EXERCISES for the
Anatomy&
Physiology
LABORATORY

Erin C. Amerman

Santa Fe Community College

Morton Publishing Company
925 W. Kenyon Ave., Unit 12
Englewood, CO 80110
http://www.morton-pub.com

Book Team

Douglas Morton	Publisher
David Ferguson	Biology Editor
Ash Street Typecrafters, Inc.	Production
Jessica Ridd	Illustration
Bob Schram, Bookends	Interior and Cover Design

For Elise

ISBN: 0-89582-658-5

Printed in the United States of America

10   9   8   7   6   5   4   3   2   1

Preface

Several years ago, when I first started teaching anatomy and physiology, my biggest frustration with the course came in the laboratory. It seemed as if I were pulling teeth to get the students interested and for them to see the connections between lecture and lab. As I attended conferences and interacted with other instructors, I found that I wasn't the only one having this problem. Many instructors were frustrated with the lab. Furthermore, they were frustrated with the lab manual. Most thought the lab manuals were too long, too expensive, and lacking in focused activities for the students. Many manuals also require expensive equipment that some colleges do not have the funds to purchase in today's era of budget cuts.

To solve these problems in my own lab, I started writing lab exercises for my students. Over the course of a few semesters, I found that the students learned from them and also actually enjoyed them because they were active for the entire two hours. Consequently, their grades on practical examinations improved dramatically.

In 2003, I was fortunate enough to be given an opportunity to expand upon my vision and share it with other instructors. The finished product contains several unique features, designed to assist the students and the instructors alike, including the following.

- *Pre-Lab Exercises*. Students who read the material prior to coming to lab tend to make better use of lab time and therefore do better on practical exams. But assigning reading prior to coming to lab is problematic for two reasons: (1) It is passive, and many students forget what they have read, and (2) many of them don't do it! I therefore incorporated Pre-Lab Exercises into each unit. The Pre-Lab Exercises are all activity-based, and include questions pertaining to the material that will be covered, as well as diagrams for the students to color-code and label. In this capacity, the Pre-Lab Exercises serve as both a study guide and as preparation for lab.

- *Organized Anatomy*. Most lab manuals have no specific lists of structures that the students are to identify. Instead, the anatomical structures are scattered throughout the unit, making it difficult for both the student and the instructor. This manual features organized lists of structures, providing a centralized list for the students, which instructors can readily customize as they wish.

- *Model Inventories*. Much of what is done in today's anatomy and physiology labs involves examining three-dimensional anatomical models. Students tend to look at one model and proclaim themselves done. But looking at one anatomical model doesn't provide students with the whole picture, nor does it allow them enough time to master the material. In response, I started using what I call "Model Inventories" in my labs. Students give the model a descriptive name, then list the structures they are able to locate on the model. This process helps them to focus more on the anatomy and to engage more of their brain as they examine, pronounce, and write down the names of the anatomical structures.

- *Focused Activities.* In addition to the model inventories, most of the units in this manual feature hands-on activities for the students. These activities were written with cost concerns in mind and seldom require special equipment or materials.

- *Tracing Exercises.* Several units feature tracing exercises, in which students follow the pathway of a substance (e.g., a molecule of glucose or an erythrocyte) throughout the body. The tracing exercises allow students to get a "big picture" of both anatomy and physiology. Although they sometimes are confused as they begin tracing, they come away with a much better understanding of the interrelationships among the systems in the body and the relationship between structure and function.

- *Affordability.* Last but not least—the price! One of the benefits of working with Morton Publishing is the staff's dedication to producing quality materials at a reasonable price. Given how expensive textbooks have become, it is nice to be able to offer these exercises to the students at a price they can afford.

Although *Exercises for the Anatomy and Physiology Laboratory* may be used alone, it also was designed to accompany *A Photographic Atlas for the Anatomy and Physiology Laboratory,* Fifth Edition, by Kent M. Van De Graaff and John L. Crawley. I hope this lab manual will provide instructors and students alike with the tools they need for a productive and interesting laboratory experience. I welcome your comments and suggestions for future editions by contacting me at compcopy@morton-pub.com.

Acknowledgments

Even though the name on the cover of this book is mine, please don't think this was a solo effort! Many people were integral to the development and production of this manual, and I'm grateful to you all.

First and foremost I would like to thank my husband Chris, my daughter Elise, and my mother Cathy. Without your unwavering support, this book would not have been possible. And to Elise, thank you especially for being patient with my being behind my computer screen so often (although you did remove the "F" key from the keyboard!). I will be eternally grateful that your first words were, "Kitty, good girl," and not, "Mommy, work, book."

Next I wish to thank the faculty at Santa Fe Community College, particularly Sarah Stone, Anna Langford, and Linda Nichols, for the fantastic ideas, suggestions, support, and experiences you provided me over the past two years. This lab manual would not have been the fine product it is without your generosity.

I also extend my gratitude to the talented book team with whom I was fortunate enough to work: Joanne Saliger at Ash Street Typecrafters, who exceeded my expectations in making the book look professional and attractive; Carolyn Acheson, who skillfully copyedited the text; and Jessica Ridd, who provided the beautiful illustrations. I thank you all for your hard work and dedication to this project.

And finally, the acknowledgments would be incomplete without thanking Doug Morton for this incredible opportunity, and Biology Editor David Ferguson, whose support, guidance, and interminable patience (I think those were your words exactly!) allowed me to make my vision a reality.

Contents

1. Introduction to Anatomical Terms

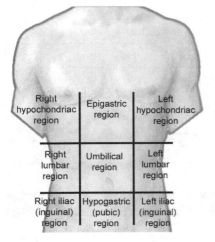

Right hypochondriac region	Epigastric region	Left hypochondriac region
Right lumbar region	Umbilical region	Left lumbar region
Right iliac (inguinal) region	Hypogastric (pubic) region	Left iliac (inguinal) region

■ OBJECTIVES

Once you have completed this unit, you should be able to:

1. Demonstrate and describe *anatomical position.*

2. Apply directional terms to descriptions of anatomical parts.

3. Use regional terms to describe various anatomical locations.

4. Demonstrate and describe anatomical planes of sections.

5. Locate and describe the divisions of the major body cavities and the membranes lining each cavity.

6. Identify organ systems, their functions, and the major organs in each system.

7. Locate and describe the functions of various organs.

■ MATERIALS

● Human torso models

● Fetal pigs (or other preserved small mammal), dissection kits, dissection trays

● Laminated outline of the human body and water-soluble marking pens

● Various anatomical models demonstrating planes of section

● Modeling clay

PRE-LAB EXERCISES

Use your text and lab atlas to complete the following exercises.

1 PRE-LAB EXERCISE 1:
Anatomical Terms

Referring to your lab atlas (Table 1.1) and text, fill in the definitions of anatomical terms in the table to the right.

2 PRE-LAB EXERCISE 2:
Organ Systems

The body has 11 organ systems, each containing certain organs and each carrying a specific subset of functions. Note, however, that some organs are part of more than one system!

In this exercise you will be identifying the 11 organ systems, their major organ(s), and the basic function(s) of each system. In addition, as you list the major organs of each system, locate these organs in your atlas and/or text. The chapters noted refer to your lab atlas.

TERM	DEFINITION
Anterior	
Posterior	
Inferior	
Superior	
Medial	
Lateral	
Proximal	
Distal	
Superficial	
Deep	
Ispilateral	
Contralateral	

ORGAN SYSTEM	MAJOR ORGANS	ORGAN SYSTEM FUNCTIONS
Integumentary System *(Chapter 4)*		
Skeletal System *(Chapter 5 and 6)*		
Muscular System *(Chapter 8)*		
Nervous System *(Chapter 9)*		
Cardiovascular System *(Chapter 12)*		
Respiratory System *(Chapter 14)*		
Lymphatic System *(Chapter 13)*		
Urinary System *(Chapter 16)*		
Digestive System *(Chapter 15)*		
Endocrine System *(Chapter 10)*		
Reproductive System *(Chapter 17)*		

EXERCISES

The bullet entered the right posterior scapular region, 3 centimeters lateral to the vertebral region, 4 centimeters inferior to the cervical region, and penetrated deep to the muscle and bone, but superficial to the parietal pleura . . .

Would you believe that by the end of this unit, you will be able to translate the above sentence and also to locate the above hypothetical wound? Unit 1 will introduce you to the world of anatomy and physiology. We will begin with an introduction to the unique language of anatomy and physiology. Like learning any new language, this may seem overwhelming at first. The key to success is repetition and application: The more you use the terms, the easier it will be to incorporate them into your normal vocabulary. From the terminology we will move on to the organization of the body into body cavities and organ systems.

Once you have completed this unit, return to the above sentence and challenge yourself to locate the precise position of the bullet wound on an anatomical model.

1 EXERCISE 1:
Anatomical Position _____

As we continue to study anatomy and physiology, most anatomical specimens are presented in a standard position termed *anatomical position*. In anatomical position, the specimen is presented facing forward, with the toes pointing forward and the feet slightly apart, and the palms facing outward (see Figure 1.1).

FIGURE 1.1
Anatomical position.

Demonstrate anatomical position for your lab partners. Use your text and the lab atlas to be certain that you are in the correct position.

✔ Check Your Understanding

Figure 1.2 is not in anatomical position. List all of the deviations from anatomical position.

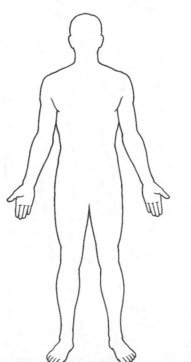

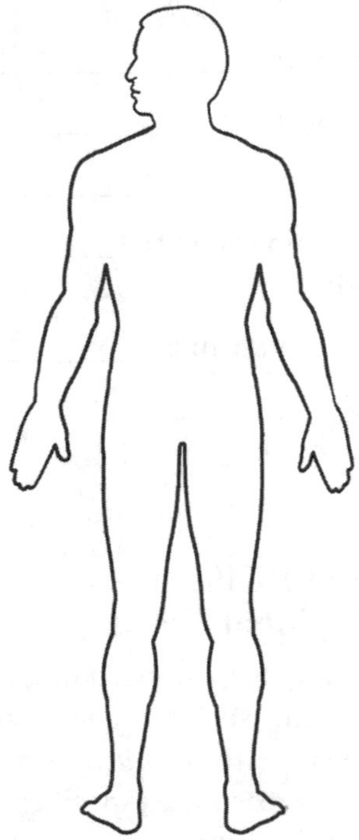

FIGURE 1.2
Figure not in anatomical position.

2 EXERCISE 2:
Directional Terms _____

In anatomy and physiology, we use certain terms to define the location of body parts and body markings. For example, when describing a wound on the chest, we could say:

● The wound is near the middle and top of the chest, or

● The wound is on the right *anterior* thoracic region, 4 centimeters *lateral* to the sternum, and 3 centimeters *inferior* to the acromial region.

The second option is precise and gives the reader a clear mental picture of the exact location of the wound.

> ➤ NOTE: WHEN WE ARE DESCRIBING THE LOCATION OF A BODY PART OR MARKING, WE ARE ALWAYS REFERRING TO A FIGURE IN ANATOMICAL POSITION.

Review the definitions of the directional terms that you completed in both of the Pre-Lab Exercises. Use these terms to fill in the correct directional term in the Practice below. As you fill in each answer, refer to diagrams such as Figure 1.2 in your atlas.

PRACTICE:
Directional Terms

Fill in the correct directional term for the items below:

1. The elbow is _____ to the wrist.

2. The chin is _____ to the nose.

3. The thumb is _____ to the forefinger.

4. The forehead is _____ to the mouth.

5. The skin is _____ to the muscle.

6. The spine is _____ to the esophagus.

7. The mouth is _____ to the ear.

8. The spine is on the _____ side of the body.

9. The right arm is _____ to the right leg.

10. The hip is _____ to the knee.

3 EXERCISE 3:
Regional Terms _____

As you may have noticed in Exercise 2 rather than using generic terms such as "chest" and "shoulder," we use the anatomical terms *thoracic region* and *acromial region*. This is done for reasons of specificity and to reduce the potential for errors in communication. "Shoulder" could consist of quite a large anatomical area, while the "acromial region" refers to one specific location on the shoulder.

Use your text and lab atlas (Figures 1.12 and 1.13) to learn the following regional terms. It looks like a long list, but you are probably familiar with several of the terms already. For example, most of us know to which regions "oral," "nasal," and "abdominal" refer. Watch for other terms that you may already know.

PRACTICE:
Labeling

As you identify each anatomical region, label it with water-soluble marking pens on the laminated outlines of the human body. Or label the regions on Figure 1.3.

Abdominal	Buccal	Femoral
Acromial	Calcaneal	Forearm
Antebrachial	Carpal	Frontal
Antecubital	Cephalic	Gluteal
Arm	Cervical	Inguinal
Axillary	Crural	Leg
Brachial	Digital	Lumbar

Mammary	Palmar	Sternal
Mental	Patellar	Sural
Nasal	Pelvic	Tarsal
Occipital	Plantar	Thigh
Oral	Popliteal	Thoracic
Orbital	Pubic	Umbilical
Otic	Scapular	Vertebral

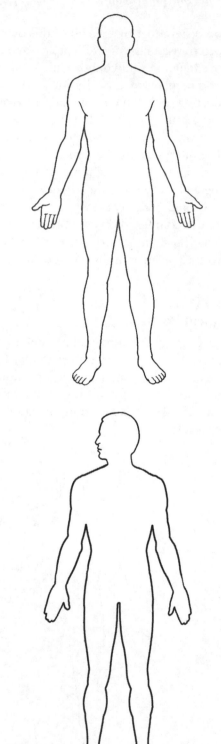

FIGURE 1.3
Anterior and posterior views of the human body in anatomical position.

✔ Check Your Understanding

You are reading a surgeon's operative report. During the course of the surgery, she made several incisions. It is your job to read her operative report and determine where the incisions were made. Draw the incisions on the laminated outlines of the human body or on Figure 1.4.

1. The first incision was made in the right anterior cervical region, 3 centimeters lateral to the trachea and 2 centimeters inferior to the mental region. The cut extended vertically to 3 centimeters superior to the thoracic region.

2. The second incision was made in the left anterior axillary region, and extended medially to the sternal region. At the sternal region the cut turned inferiorly to 4 centimeters superior to the umbilical region.

3. The third incision was made in the left posterior scapular region. The cut was extended medially to 2 centimeters lateral to the vertebral region, where it turned superiorly and progressed to 1 centimeter inferior to the cervical region.

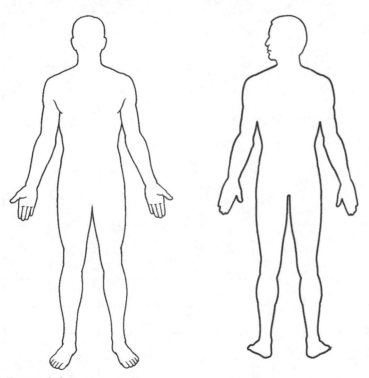

FIGURE 1.4
Anterior and posterior views of the human body in anatomical position.

4 EXERCISE 4: Body Cavities and Membranes

The body is divided into several fluid-filled cavities, each containing specific organs. In this exercise you will be identifying the body cavities, as well as the organs contained within each cavity. Refer to Figures 1.8, 1.9, and 1.10 in your lab atlas as you read about the cavities.

There are two major subdivisions, each subdivided into smaller cavities, as follows:

1. *Dorsal (posterior) cavity*. It contains two smaller cavities:

   a. *Cranial cavity*: the area encased by the skull.

   b. *Vertebral* (or *spinal*) *cavity*: the area encased by the vertebrae.

2. *Ventral (anterior) cavity*. It contains two smaller cavities, each of which can be further subdivided into still smaller cavities. The divisions of the ventral cavity are filled with a watery fluid called *serous fluid*, and are surrounded by membranes called *serous membranes*. The serous membranes will be discussed later in this exercise.

   a. *Thoracic cavity*: the area superior to the diaphragm and encased by the ribs. It may be divided into:

      (1) *Pleural cavities*: the area surrounding the lungs. There are two pleural cavities.

      (2) *Mediastinum*: the area between the pleural cavities. It has one smaller cavity housed within it:

         (a) *Pericardial cavity*: the cavity that surrounds the heart.

   b. *Abdominopelvic cavity*: the area inferior to the diaphragm, extending to the bony pelvis. It contains two smaller cavities:

      (1) *Abdominal cavity*: the area superior to the bony pelvis.

      (2) *Pelvic cavity*: the cavity housed within the bony pelvis.

In addition to the smaller cavities, the abdominopelvic cavity may be divided into four quadrants or nine regions based upon a series of lines drawn through the surface of the cavity. The regions are listed and labeled in Figure 1.5.

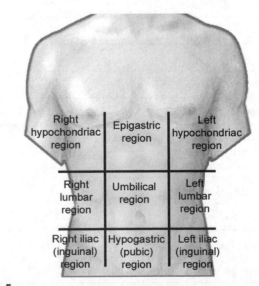

FIGURE 1.5
Regions of the abdominopelvic cavity.

Body Cavities

To visualize the cavities, you will either use a human torso model or a preserved small mammal such as a cat, a fetal pig, or a rat. Instructions for opening all three animals may be found in your atlas in Chapter 19 (cat), 20 (fetal pig), and 21 (rat).

> ➤ NOTE: SAFETY GLASSES AND GLOVES ARE REQUIRED!

As you open the fetal pig or human torso model, pay close attention to the divisions between cavities. Use Figures 1.8–1.11 in your lab atlas to assist you. As you locate and identify each cavity, do the following:

1. List the organs you are able to see in the chart below.

2. When examining the abdominopelvic cavity, mark each region with a pin or marking tape (if working with a torso). Note which specific organs are visible in each region.

Serous Membranes

The ventral body cavities are surrounded by sheets of tissue called *serous membranes*, which secrete a thin, watery fluid called *serous fluid*. Serous fluid lubricates the organs in the cavity, allowing them to contract and relax and slide past one another without friction.

Serous membranes are composed of two layers, the outer *parietal* layer, which is attached to the body wall, and the inner *visceral* layer, which is attached to the organ or organs. Between the parietal and visceral layers lie the cavities, which are filled with serous fluid. For example, the pleural cavity is the space between the parietal pleural membrane and the visceral pleural membrane.

Serous membranes are best visualized on a preserved specimen such as the fetal pig, as it is difficult to appreciate their structure on a model. As you dissect the fetal pig, look for the serous membranes listed in the accompanying chart.

CAVITY	ORGAN(S)
Dorsal cavity:	
1. Cranial cavity	
2. Vertebral cavity	
Ventral cavity:	
1. Thoracic cavity	
a. Pleural cavities	
b. Mediastinum	
(1) Pericardial cavity	
2. Abdominopelvic cavity	
a. Subdivisions:	
(1) Abdominal cavity	
(2) Pelvic cavity	
b. Regions:	
(1) Right hypochondriac region	
(2) Epigastric region	
(3) Left hypochondriac region	
(4) Right lumbar region	
(5) Umbilical region	
(6) Left lumbar region	
(7) Right iliac region	
(8) Hypogastric region	
(9) Left iliac region	

Take care not to tear the fragile membranes, which consist of little more than a few layers of cells. As you identify each membrane, state in which cavity the membrane is found, and the structure on which the membrane rests (the lungs, heart, abdominal wall, etc.). Use that information to fill in the table below.

Use either Figure 1.6 or your laminated outline of the human body to draw in the body cavities using either colored pencils or different-colored water-soluble marking pens. Also, where applicable, state which membranes are found surrounding the cavity.

MEMBRANE	CAVITY	STRUCTURE
Parietal pleura		
Visceral pleura		
Parietal pericardium		
Visceral pericardium		
Parietal peritoneum		
Visceral peritoneum		

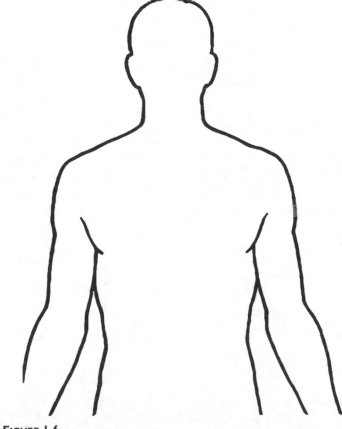

FIGURE 1.6
Anterior view of the human torso in anatomical position.

✔ Check Your Understanding

You are acting as coroner, and you have a victim with three gunshot wounds. In this scenario, your "victim" will be your fetal pig or a human torso model that your instructor has "shot." State the anatomical region and/or body cavity in which the "bullet" was found, and describe the location of the wound using at least three directional terms. As coroner, you have to be as specific as possible and keep your patient in anatomical position!

Shot 1	
Shot 2	
Shot 3	

5 EXERCISE 5: Planes of Section

Often in science and in the medical field, it becomes necessary to obtain different views of the internal anatomy of an organ or a body cavity. To obtain these views, it is useful to make an anatomical section along a specific plane. The four commonly used planes of section are:

1. *Sagittal plane.* A section along the sagittal plane divides the body part into right and left halves. The two variations of the sagittal section are:

   a. *Midsagittal* sections divide the body part into equal right and left halves.

   b. *Parasagittal* sections divide the body part into unequal right and left halves.

2. *Frontal plane.* Also known as the *coronal plane*, this plane of section divides the body part into an anterior (front) part and a posterior (back) part.

3. *Transverse plane.* Also known as a *cross-section*, this plane of section divides the body part into a superior (or proximal) part and an inferior (or distal) part.

4. *Oblique section*: This section cuts at an angle and is intended to visualize structures that are difficult to see with standard angles.

Visualizing Anatomical Planes

Use a scalpel to cut a ball of modeling clay in each of the following anatomical planes:

1. Sagittal
   a. Midsagittal
   b. Parasagittal
2. Frontal
3. Transverse
4. Oblique

See Figures 1.5 and 1.7 in your lab atlas for reference.

> ➤ NOTE: IT IS USEFUL TO MOLD YOUR CLAY INTO THE SHAPE OF A HEAD AND TO DRAW EYES ON THE HEAD TO DENOTE ANTERIOR AND POSTERIOR SIDES.

Identifying Anatomical Planes

Using the models your instructor provides, identify at least two examples of each plane of section listed above, and state which organs are visible in the section. Record your answers in the chart below.

EXAMPLES OF MIDSAGITTAL AND/OR PARASAGITTAL SECTIONS:	
Model:	**Organ(s) visible:**
1.	
2.	

EXAMPLES OF FRONTAL SECTIONS:	
Model:	**Organ(s) visible:**
1.	
2.	

EXAMPLES OF TRANSVERSE SECTIONS:	
Model:	**Organ(s) visible:**
1.	
2.	

✔ Check Your Understanding

1. What may be some real-world applications of anatomical sections? (*Hint*: Think of the medical field.)

2. Which anatomical section(s) would provide a view of the internal anatomy of *both* kidneys?

3. If you were to make a transverse section straight through the diaphragm (not superior or inferior to it), through which body cavity would you cut? (*Hint*: Don't forget about the dorsal body cavities.)

6 EXERCISE 6:
Organs and Organ Systems _____

In the Pre-Lab Exercises, you listed the 11 organ systems, as well as the primary organs of each system and their functions. In this exercise you will apply this knowledge to identify the major anatomical structures of each organ system.

Organs

To familiarize yourself with the major organs in the body, as well as the organs' textures, identify the following organs on your fetal pig or human torso models. Check off each organ as you identify it, using Chapter 20 in your lab atlas for reference.

Brain	Liver
Spinal cord	Intestines
Heart	Gallbladder
Lungs	Pancreas
Larynx	Adrenal glands
Trachea	Thyroid glands
Kidneys	Testes or ovaries
Urinary bladder	Spleen
Esophagus	Thymus
Stomach	

Representative Models of Organ Systems

Your instructor has selected various models to represent the different organ systems of the body. For this exercise, figure out which organ system each model is intended to represent.

ORGAN SYSTEM	MODEL(S) REPRESENTING
Integumentary system	
Skeletal system	
Muscular system	
Nervous system	
Cardiovascular system	
Lymphatic system	
Respiratory system	
Digestive system	
Urinary system	
Endocrine system	
Reproductive system	

✔ Check Your Understanding

The type of anatomy we are studying in this lab manual is called *systemic anatomy,* meaning that we cover all organs related to a specific system. Some, however, choose to study anatomy from a regional point of view (e.g., the abdominal region or the thoracic region). Find at least two organ systems that contain organs in different regions of the body. Also, state which organs are located outside of the main region for that system.

Example: The nervous system has organs in the cranial cavity (the brain), the spinal cavity (the spinal cord), the thoracic and abdominopelvic cavities (spinal and cranial nerves), and the extremities (spinal nerves).

2. Chemistry

■ OBJECTIVES

Once you have completed this unit, you should be able to:

1. Demonstrate the proper interpretation of pH paper.
2. Apply the pH scale.
3. Understand the purpose and effects of a buffer.
4. Apply laboratory techniques to perform aspirin synthesis and DNA extraction.

■ MATERIALS

- pH paper
- Samples of various acids and bases
- 0.1M Hydrochloric acid
- Samples of various antacid tablets
- Buffer solution
- 0.2M NaOH
- Willow bark
- Ethanol
- Glacial acetic acid
- Ferric chloride
- Pipettes and spot plates
- Pea soup mixture
- Isopropyl alcohol
- Enzymes (meat tenderizer)
- 50 mL beakers
- Glass stirring rods
- Test tubes
- Liquid detergent

PRE-LAB EXERCISES

Prior to coming to lab, use your text and lab atlas to complete the following exercises.

PRE-LAB EXERCISE 1: Chemical Bonding _____

One of the more difficult concepts for students to grasp in chemistry is chemical bonding. Unfortunately, bonding is something that pops up again and again and again. We revisit it in cytology, the nervous system, blood, respiration, and digestion, to name just a few.

In this exercise, use the chemistry chapter of your textbook to define:

1. Metal:

   Do metals like to donate or accept electrons?

2. Nonmetal:

   Do nonmetals like to donate or accept electrons?

3. Electron shells:

   How many electrons go in the first shell? _____
   How many electrons go in the second shell? _____
   How many electrons go in the third shell? _____

4. Valence electrons:

   How many electrons does each element want in its valence shell? _____

5. Anion:

6. Cation:

Now define, according to your text, the following types of chemical bonds:

1. Ionic bonds:

2. Covalent bonds:

   a. Polar covalent bonds:

   b. Nonpolar covalent bonds:

Okay, now that we have technical definitions for all of these terms, let's try to decipher what it all means:

Example 1: Sodium and Chlorine

Let's take a look at sodium (Na) and chlorine (Cl). Draw what I am describing, and you will see it better.

Sodium has how many protons? _____

Therefore, we know that Na has how many electrons? _____

How many electrons will go in the first shell? _____

How many in the second shell? _____

How many in the third? _____

Now draw this out on the diagram below, and take a look at it, in particular the third (valence) shell. Note that the black represents the nucleus of the atom.

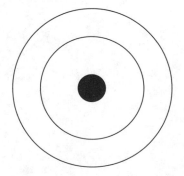

We know that Na wants eight electrons in its valence shell. But how many does it have? _____

So to fill this shell, is it going to be easier for sodium to steal seven more electrons from another atom, or would it just be easier for Na to give up that one electron and get rid of that third shell? Na is simply going to *give away* that last electron. This means it will lose an electron (negative charge), but will keep the same number of protons (positive charges). What will sodium's overall charge be now? _____

Now let's look at chlorine:

Chlorine has how many protons? _____

Therefore, we know that Cl has how many electrons? _____

How many electrons will go in the first shell? _____

How many in the second shell? _____

How many in the third? _____

Again, draw this out:

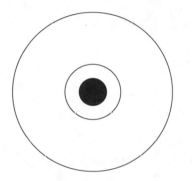

We know that Cl wants eight electrons in its valence shell. Is it going to be easier for Cl to steal one electron from another atom, or would it be easier for Cl to give away seven electrons? It will be *much* easier for chlorine to take an electron from another atom than to try and pawn off seven electrons. This means that chlorine will gain an electron (negative charge) but will keep the same number of protons (positive charges). What charge will chlorine (now called *chloride*) have now? _____

Let's tie this in with what we learned about metals and nonmetals earlier.

We said that metals like to _____ electrons, to become _____.

Is sodium a metal or a nonmetal? _____

Looking at our example above, did sodium donate or accept an electron? _____

We also said that nonmetals like to _____ electrons, to become _____.

Is chlorine a metal or a nonmetal? _____

Looking at our example above, did chlorine donate or accept an electron? _____

Why do we care? Well, when sodium donates an electron to become a positively charged cation, and chlorine accepts an electron to become a negatively charged anion, this creates an attraction. (Think of it like "opposites attract.") This attraction leads the two ions to *bond* to one another, forming an *ionic bond*. See how simple that is?

Example 2: Polar Covalent Bond

Now let's tackle another difficult bonding concept, the *polar covalent bond*. We established earlier that the polar covalent bond results from the unequal sharing of electrons. Essentially, one atom is "greedier" than the other, and "hogs" the electrons. This creates what is known as a dipole—a molecule with a partial positive pole and a partial negative pole ("dipole" literally meaning "two poles"). The classic example of a polar covalent bond is water:

Hydrogen (H) has how many protons? _____

Therefore, we know that H has how many electrons? _____

How many electrons will go in the first shell? _____

How many electrons would H need to fill up that first shell? _____

Oxygen (O) has how many protons? _____

Therefore, we know that O has how many electrons? _____

How many electrons will go in the first shell? _____

How many electrons will go in the second shell? _____

How many electrons would O need to fill up that second shell? _____

Now, in this case, rather than either element giving up or accepting electrons, they share them instead. Because O needs two electrons to fill that valence shell, and H only has one electron to share, O needs to form a partnership with two hydrogen atoms, forming H_2O. But O is the greedier atom, meaning that the electrons will spend more time around the oxygen, and less around the hydrogen. Thus, a polar covalent bond is born. This means that:

O will have a partial _____ charge, and the

two hydrogens will have a partial _____ charge.

Keep in mind that a nonpolar covalent bond involves the same principles, except that the atoms share the electrons equally. As a result, there is no dipole because the electrons don't spend any more time around one or more of the atoms.

Following are some simple rules for determining whether a compound has polar covalent, nonpolar covalent, or ionic bonds.

- Any metal bound to a nonmetal has ionic bonds.
- All compounds with only C-H bonds are nonpolar covalent.
- Any compound containing only identical atoms (e.g., H_2) has nonpolar covalent bonds.
- Any nonmetal bound to O or N has polar covalent bonds.
- Any compound containing mostly C and H has non-polar covalent bonds.

PRACTICE

Practice what you have learned by filling in the required information in the table below.

2 PRE-LAB EXERCISE 2: Acids, Bases, and pH _____

In this exercise, we will examine the pH of several substances. Prior to coming to lab, answer the following questions pertaining to acids, bases, and the pH scale.

Define the following:

1. Acid:

2. Base:

3. Buffer:

4. pH scale:

a. What is the pH of an acid?

b. What is the pH of a base?

c. Which pH is considered neutral?

Substance	Molecular Formula (or atomic symbol)	Element or Compound	Metal or Nonmetal	Type of Bond (or type of bonding that may occur if the substance is an element)
Potassium				
Water				
Sodium chloride				
Carbon				
Glucose				

EXERCISES

In the following exercises you will apply chemistry principles to determine pH and see the effects of buffers on the pH of solutions. Then, applying these concepts, you will synthesize aspirin and extract DNA. It's not every day that you get to say that you made aspirin and extracted DNA, is it?

1 EXERCISE 1: Determining pH

The *pH* is a common measurement in science and medicine. It is a measure of the concentration of hydrogen ions present in a solution. As the hydrogen ion concentration increases, the solution becomes more *acidic*. As the hydrogen ion concentration decreases, the solution becomes more *alkaline*, or *basic*.

A simple way to measure pH is to use pH paper. To test the pH with pH paper, drop one or two drops of the sample solution on the paper and compare the color change with the colors on the side of the pH paper container. The pH is read as the number corresponding to the color that the paper turned.

Reading the pH

> NOTE: SAFETY GLASSES AND GLOVES ARE REQUIRED.

Obtain two samples each of acids and bases, and measure their pH using pH paper. Record the pH values in the table below. Once you have tested four known samples, test two randomly selected unknown samples in the same manner with pH paper.

SAMPLES	MOLECULAR FORMULA	PH
Acid #1		
Acid #2		
Base #1		
Base #2		

UNKNOWN SAMPLES	PH	ACID OR BASE?
Unknown #1		
Unknown #2		

pH Applications

> NOTE: SAFETY GLASSES AND GLOVES ARE REQUIRED.

Now let's apply the pH scale to physiologic systems. The stomach contains concentrated hydrochloric acid, and the pH of the stomach contents ranges between 1 and 3. Antacids are a group of medications designed to neutralize stomach acid to treat a variety of conditions, including gastroesophageal reflux, commonly known as heartburn. Antacids generally consist of a metal bound ionically to a base. For this application you will be comparing the effectiveness of three widely available antacids in neutralizing concentrated hydrochloric acid.

Procedure

1. Obtain three glass test tubes, and label them tubes 1, 2, and 3.
2. Obtain a bottle of 0.1 M hydrochloric acid (HCl), and record its pH in the blank below.

   pH of 0.1 M HCl: _____

3. In each tube, place 2 mL of the 0.1 M HCl.
4. To tube 1, add one-fourth of one Tums® tablet that has been crushed.
5. To tube 2, add one-fourth of one Rolaids® tablet that has been crushed.
6. To tube 3, add one-fourth of one Alka-Seltzer® tablet that has been crushed.
7. Allow the tubes to sit undisturbed for 3 minutes.
8. Measure the pH of the contents of each tube, and record the values in the table below.

ANTACID	ACTIVE INGREDIENT	PH
Tums®		
Rolaids®		
Alka-Seltzer®		

9. Based upon your observations, which antacid is the most effective?

✔ Check Your Understanding

1. We have seen that the pH of the stomach is very acidic (around 1–3). Most tissues at this pH would be severely damaged. Use your text to discover how the stomach prevents itself from becoming damaged.

2. When stomach acid bubbles up into the esophagus, a condition called gastroesophageal reflux, it tends to produce a burning, painful feeling. Why does the acid hurt the esophagus but not the stomach?

2 EXERCISE 2: Buffers

In the Pre-Lab Exercises, you defined a buffer as a chemical that resists changes in pH. In this exercise, you will examine the effects of adding acid or base to buffered solutions and non-buffered solutions.

Procedure

1. Label four wells in a plate as 1, 2, 3, and 4.
2. Fill wells 1 and 2 about half-full of distilled water; this is the non-buffered solution. Measure the pH of the distilled water and record that value in the table below ("pH before").
3. Fill wells 3 and 4 about half-full of buffered solution. Measure the pH of the buffered solution and record that value in the table below ("pH before").
4. To wells 1 and 3, add two drops of 0.1 M HCl. Stir the solutions and measure the pH of each well. Record the pH in the table below ("pH after").
5. To wells 2 and 4, add two drops of 0.2 M NaOH (a base). Stir the solutions and measure the pH of each well. Record the pH in the table below ("pH after").

WELL	CONTENTS	pH BEFORE	pH AFTER
1	Water and HCl		
2	Water and NaOH		
3	Buffer and HCl		
4	Buffer and NaOH		

✔ Check Your Understanding

1. What is the pH for human blood?

2. To maintain the pH of the blood within this narrow range, the body has a system of buffers to resist sudden changes in pH. What are the primary buffer systems used in the human body?

3. Use your text to determine what happens if the buffer systems are overwhelmed and:

   a. The blood becomes excessively basic (a condition called alkalosis):

   b. The blood becomes excessively acidic (called acidosis):

4. What could cause the above conditions?

3 EXERCISE 3: Aspirin Synthesis

In this exercise, we will make one of the most commonly used over-the-counter medications, *aspirin*, an anti-inflammatory drug. Aspirin is derived from the chemical *salicylic acid*, which was originally used in the treatment of *inflammation*, much as aspirin is today. Salicylic acid, however, contains a chemical group called a *phenol*, which makes salicylic acid irritating to the stomach. Fortunately, by removing the phenol group and replacing it with an acetic acid (vinegar) group, the drug is less irritating to the stomach.

To synthesize aspirin (chemical name acetylsalicylic acid, or ASA), we first must isolate the chemical salicylic acid from its source, willow bark. This is done by soaking the willow bark in ethanol. Then we must *acetylate* the salicylic acid by adding acetic acid to it, which removes the phenol group and replaces it with an acetic acid group.

Procedure

> NOTE: SAFETY GLASSES AND GLOVES ARE REQUIRED.

1. Fill the large well with willow bark, and cover the bark with ethanol.

2. Soak the bark in ethanol for 15 minutes.

3. Label two smaller wells as well 1 and well 2.

4. After 15 minutes, use a pipette to remove several drops of the ethanol covering the willow bark, which now contains the extracted salicylic acid, and place equal amounts into well 1 and well 2.

5. Add two drops of iron chloride (also called ferric chloride) into well 1. Iron chloride reacts with phenol groups to turn a dark purple-brown color. If no phenol groups are present, the solution remains orange.

   Color of solution in well 1: _____

   Does the solution in well 1 contain salicylic acid? How do you know?

6. Add 15 drops of acetic acid (vinegar) to the solution in well 2, and stir for 15 seconds. You have now synthesized aspirin!

7. Into well 2, add two drops of iron chloride.

   Color of solution in well 3: _____

   Was your synthesis of aspirin successful? How can you tell?

✔ Check Your Understanding

1. There has been a recent surge in the popularity of herbal medical remedies. Most of these remedies claim that, because they are derived from plants, they are "all natural" and "drug-free." Given the synthesis you just performed, what do you think of such claims?

2. Why do you think aspirin (acetylsalicylic acid), which has a pH of about 6, is less irritating to the stomach than salicylic acid, which has a pH of about 3?

4 EXERCISE 4: DNA Extraction _____

In this exercise, we will be extracting *deoxyribonucleic acid* (DNA) from dried green peas. Although this is obviously not animal DNA, keep in mind that it is composed of precisely the same nucleic acid bases as animal DNA, and that the plant cell membrane structure is similar to that of animal cells.

Much of what we will do in this extraction relies on our knowledge of the bonding principles discussed in the Pre-Lab Exercises. To separate chemicals, we can take advantage of their differing *solubilities* in solutions with polar covalent and nonpolar covalent bonds. In general, the following rule can be used to determine the solubility of one compound in another compound:

Like dissolves like.

This means that chemicals with polar covalent bonds prefer to interact with chemicals that also have polar covalent bonds. Chemicals with nonpolar covalent bonds prefer to interact with chemicals that also have nonpolar covalent bonds. Think about what happens when you combine oil (nonpolar covalent compound) and water (polar covalent compound). Do they mix? In their natural condition, oil and water don't mix, because of their differing types of chemical bonds.

Not all chemicals have only polar or nonpolar bonds. Certain chemicals are *amphipathic*, and have both polar and nonpolar parts. In class you may have discussed the cell membrane, which is composed of a bilayer of amphipathic chemicals called *phospholipids*.

Procedure

> ➤ NOTE: SAFETY GLASSES AND GLOVES ARE REQUIRED.

1. Obtain about 10 mL of the blended pea soup mixture (which contains only dried peas and water), and pour it into a 50 mL beaker.

2. Add 1.6 mL of liquid detergent to the pea soup. Stir gently with a glass stirring rod, and let the mixture sit undisturbed for 5–10 minutes.

3. After 5–10 minutes, pour the mixture into a test tube, filling it about one-third full.

4. Add a pinch of meat tenderizer (which contains enzymes) to the test tube and stir with a glass stirring rod *very gently* (rough stirring will break up the DNA, making it harder to visualize).

5. Tilt the test tube and slowly add isopropyl alcohol so the alcohol forms a layer on the top of the pea mixture. The amount of alcohol you add should be approximately equal to the volume of the pea mixture.

6. Watch as the DNA (it looks like long, stringy stuff) rises to the alcohol layer.

7. Using an applicator stick, fish out the DNA and set it on a paper towel. Voila! You have just extracted DNA!

✔ Check Your Understanding

1. Detergents have a structure very similar to the phospholipids in the cell membrane. Knowing this, why do we use detergents in the first step of the extraction? (*Hint*: Think about the rule that "like dissolves like.")

2. The enzymes in meat tenderizer are called *proteases*. What do proteases degrade? Why would we need proteases in this extraction? (*Hint*: Think about the structure of DNA. What is in it besides nucleic acids?)

3. Does the isopropyl alcohol mix with the water/pea mixture? What does this tell you about the type of bonds (ionic, polar covalent, or nonpolar covalent) in isopropyl alcohol?

4. Did the DNA prefer to interact with the water/pea mixture or with the alcohol? What does this tell you about the type of bonds present in DNA?

3. Introduction to the Microscope

■ **OBJECTIVES**

Once you have completed this unit, you should be able to:

1. Identify the major parts of the microscope.
2. Define the magnification of high, medium, and low power.
3. Demonstrate proper use of the microscope.
4. Practice focusing on low, medium, and high power using introductory slides.
5. Define depth of focus.

■ **MATERIALS**

- Light microscopes with three powers
- Oil immersion microscope for demonstration
- Oil
- Lens paper
- Introductory slides (three colored threads and the letter 'e')

EXERCISES

Working with the microscope and slides seems to be one of the least favorite tasks of anatomy and physiology students. But with a bit of help, a fair amount of patience, and a lot of practice, the use of microscopes becomes progressively easier with each unit (and, yes, we do look at slides in almost every unit!).

1 EXERCISE 1:
Introduction to the Microscope _____

The microscopes that you will use in this lab are called *light microscopes*. This type of microscope shines light through the specimens to illuminate them, and the light is refracted through objective lenses to magnify them. Light microscopes have the following components:

- *Ocular lens*: the lens through which you look to examine the slide. The microscope may have one ocular lens (a *monocular* microscope) or two ocular lenses (a *binocular* microscope). Many ocular lenses have pointers that can be moved by rotating the black *eyepiece*.

- *Objective lenses*: lenses of various powers of magnification. Most microscopes have low (4X), medium (10X), and high (40X) power objective lenses. The objective lenses are attached to the *nosepiece*, which allows the operator to switch between objectives. Please note that certain microscopes have a higher power objective

(100X), called the *oil immersion lens*, which requires a drop of oil to be placed between the slide and objective lens. If available, your instructor may have an oil immersion microscope set up for demonstration purposes.

- *Stage*: the surface on which the slide sits. It typically has stage clips that hold the slide in place. The stage on many microscopes is moveable using the mechanical stage adjustment knob. Others require you to move the slide manually.

- *Arm*: supports the body of the microscope and typically houses the adjustment knobs.

- *Coarse adjustment knob*: a large knob on the side of the arm that moves the stage up and down to change the distance of the stage from the objective lenses. It allows gross focusing of the image.

- *Fine adjustment knob*: a smaller knob that allows fine tuning of the image's focus.

- *Lamp* (also called the illuminator): provides the light source. It rests on the *base* of the microscope.

- *Iris diaphragm*: an adjustable wheel on the underside of the stage that controls the amount of light that is allowed to pass through the slide.

Microscope Parts

Label Figure 3.1 with the parts of the light microscope.

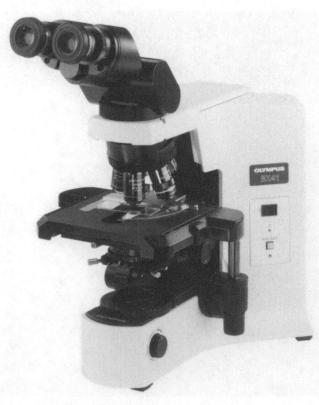

FIGURE 3.1
Parts of the light microscope.

Magnification

As discussed, to obtain magnification, light is refracted through two lenses: the ocular lens and the objective lens. Magnification of the ocular lens is always 10X (meaning that it is magnified 10 times). Magnification of the objective lens varies, but typically is 4X for low power, 10X for medium power, and 40X for high power. (To verify that this is the case for your microscope, look at the side of the objective lens, which is usually labeled with its magnification.) Remember that oil immersion provides even greater magnification at 100X. The total magnification is obtained by multiplying the power of the ocular lens by the power of the objective lens.

Fill in the following chart to determine total magnification at each power:

POWER	MAGNIFICATION OF OCULAR LENS	MAGNIFICATION OF OBJECTIVE LENS	TOTAL MAGNIFICATION
Low			
Medium			
High			

2 EXERCISE 2: Focusing and Using the Microscope _____

As I'm certain your instructor will point out, microscopes are expensive! Running around $500–$1500 each, care must be taken to ensure that the microscopes stay in good working condition. Taking proper care of microscopes makes the histology sections of your labs run more smoothly, and it also ensures that you stay on your lab instructor's good side! Bearing that in mind, following are some general guidelines for handling the microscopes:

● When carrying the microscope, use two hands to support it: one hand holding the arm and the other supporting the base.

● After plugging in the microscope, gather up the cord so that it does not dangle off of the lab table. This will help to prevent accident-prone people (like myself!) from tripping over loose cords.

● Clean the lenses with lens paper only. Do not use paper towels or cloth to clean the lenses, as this will result in scratched lenses.

● Before you begin, make sure the nosepiece is switched to the lowest power objective.

● Get the image in focus on low power, then switch to higher power and adjust with the fine adjustment knob. Be careful not to use the coarse adjustment knob with the high-power objective in place, as you could break the slide and damage the lens.

If you follow these general guidelines, you can rest assured that the microscopes (and your grade) will not suffer any harm.

Focusing the Microscope

Now that we know how to properly handle the microscope, let's practice using it.

1. Obtain a set of practice slides, with a slide of the letter "e."

2. Examine the letter e slide before placing it on the stage. How is the "e" oriented on the slide? Is it right side up, upside down, backwards, etc.?

3. Ensure that the nosepiece is switched to low power, and place the slide on the stage, securing it with the stage clips.

4. Looking through the ocular lens, use the coarse adjustment knob to slowly bring the slide into focus. Once it is grossly in focus, you may use the fine adjustment knob to sharpen the focus. How is the "e" oriented in the field of view? Is it different than it was when you examined it in item 2?

5. Move the nosepiece to medium power. You should have to adjust the focus only with the fine adjustment knob; no adjustment of the coarse focus should be necessary.

6. Once you have examined the slide on medium power, move the nosepiece to high power. Again, focus only with the fine adjustment knob.

Wasn't that easy?

Depth of Focus

Frequently, in my introductory microscopy laboratories, students call me over to their table to look at something they have found on the slide. They think that what they found is the specimen, when what they actually found is dirt on the top of the slide. This is so because students tend to focus the objective on the first thing they can make out, which usually is the top of the coverslip on the slide, which has a tendency to be dirty.

Why does this happen so often? It is difficult to get used to what is known as the *depth of focus*, the thickness of a specimen that is in sharp focus. Certain thicker specimens will require that you focus up and down to look at all levels of the specimen. This takes practice and skill.

1. Obtain a slide with three colored threads. The threads have been placed on the slide at varying depths, and you will have to focus on each thread individually.

2. Examine the slide prior to putting it on the stage.

3. Ensure that the nosepiece is switched to low power, and place the slide on the stage, securing it with the stage clips.

4. Use the coarse adjustment knob to get the slide into focus on low power.

5. Switch to medium power, and use the fine adjustment knob to sharpen the focus. Which thread(s) is(are) in focus?

6. Move the objective up and down slowly with the coarse adjustment knob, focusing on each individual thread. Figure out which color thread is on the bottom, in the middle, and on top, and write the color below:

Bottom _____

Middle _____

Top _____

Hints for Using the Microscope

Okay, so you can focus on newsprint and threads, but what about cellular structures and tissue sections? Well, those are certainly more difficult, but if you keep the following hints in mind, the task becomes much less daunting:

- *Always start on low power.* You are *supposed* to start on low power anyway, to avoid damaging the objective lenses. Sometimes, though, students forget and jump straight to medium or high power. This risks damaging the lenses and also makes it harder on you. Bear in mind that most slides will have more than one histological or cellular structure on each slide. Starting on low power allows you to scroll through a large area of the slide and then focus in on the desired part of the section.

- *Beware of too much light.* It is easy to wash out the specimen with too much light. If you are having difficulty making out details, use the iris diaphragm to reduce the amount of light illuminating the specimen. This will increase the contrast and allow you to observe more details.

- *Keep both eyes open.* When looking through a monocular microscope, it is tempting to close one eye. Admittedly, keeping both eyes open is difficult at first, but it will help prevent eyestrain and headaches.

- *Compare your specimen to the photos in your atlas.* Although the slides you are examining will not necessarily be identical to the photos in your atlas, they should be similar in appearance. Generally speaking,

if you are looking at something that is vastly different from what is in the atlas, you probably should move the slide around a bit to find the correct tissue or cell type on the slide.

- *Remember that the slides aren't perfect.* Not all slides will clearly demonstrate what you need to see. Some aren't stained adequately or properly. Some are sectioned at a funny angle. Some don't contain all of the tissue or cell types you need to see. Most will not look as nice as they do in the atlas. What should you do about this? See the next hint for the answer.

- *Look at more than one slide of each specimen.* This will help you in the face of sub-par slides and also will assist you overall in gaining a better understanding of the specimens you are examining.

- *Draw what you see.* Students tend to resist drawing, often claiming artistic incompetence. But even the most basic picture is helpful for two reasons. First, it allows you to engage more parts of your brain in the learning process. The more areas of your brain that are engaged, the better the chances that you will retain the information. Also, drawing is helpful in that it means you actually have to look at the specimen long enough to draw it!

- *Have patience!* It really does get easier. Don't get frustrated, and don't give up. By the end of the semester, you may come to appreciate the microscope and the fascinating world it reveals!

4. Cytology

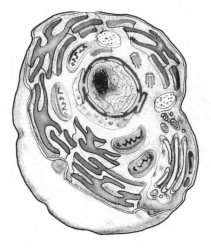

■ **OBJECTIVES**

Once you have completed this unit, you should be able to:

1. Identify parts of the cell and organelles on models, diagrams, and microscope slides.

2. Observe the effects of hypotonic, isotonic, and hypertonic environments on red blood cells.

3. Prepare and observe fetal pig cheek cells.

4. Identify the stages of the cell cycle and mitosis.

■ **MATERIALS**

- Cell models
- Colored pencils
- Cell slides (with red blood cells, skeletal muscle, and sperm cells)
- Animal or human blood cells
- Blank slides and coverslips
- 5% dextrose in water solution
- Deionized water
- 25% NaCl in water solution
- Fetal pigs
- Wooden applicator sticks
- Methylene blue dye
- Mitosis models and slides

PRE-LAB EXERCISES

Prior to coming to lab, use your text and lab atlas to complete the following exercises.

1 PRE-LAB EXERCISE 1:
Organelles

The following is a list of organelles we will identify in lab. Use your text and lab atlas to determine the functions of these organelles.

ORGANELLE	FUNCTION
Cell membrane	
Nucleus	
Nuclear membrane	
Nucleoli	
Cytoplasm (cytosol)	
Ribosomes	
Smooth endoplasmic reticulum	
Rough endoplasmic reticulum	
Golgi apparatus	
Lysosomes	
Peroxisomes	
Mitochondria	
Centrioles	
Microtubules and microfilaments	
Vesicle	
Microvilli	
Cilia	
Flagella	

2 PRE-LAB EXERCISE 2:
Components of Cell Membrane

Figure 4.1 is a diagram of the cell membrane. Label the components of the cell membrane, using the terms from Pre-Lab Exercise 1. In addition, color-code the various components, coloring the nonpolar portions of the phospholipids and the proteins a different color from the polar portions of these molecules. See Figure 2.1 in your lab atlas for reference.

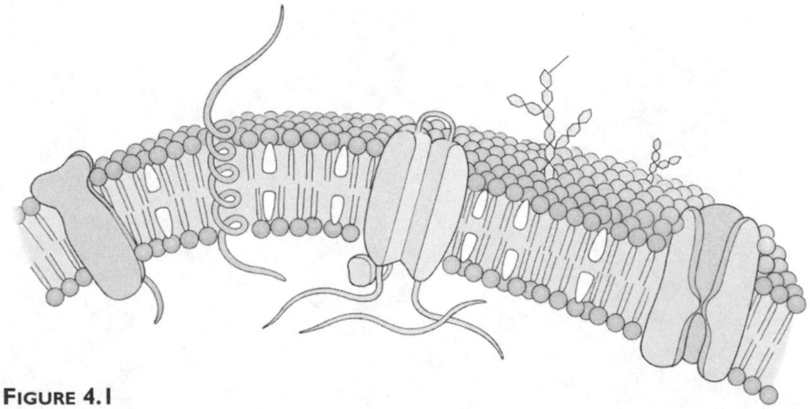

FIGURE 4.1
Phospholipid bilayer of cell membrane.

3 PRE-LAB EXERCISE 3:
Components of a Cell_____

Figure 4.2 is a diagram of the cell. Label the components of the cell with the terms from Pre-Lab Exercise 1 (not all structures may be visible in this diagram). You also may use colored pencils to color-code the different structures of the cell. (It may be helpful to color-code the cell structures in a similar manner as the cell models you use in lab.) See Figures 2.1 and 2.2 in your lab atlas for reference.

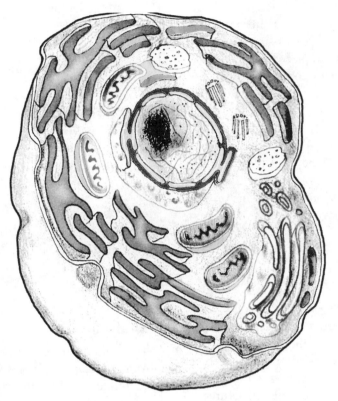

FIGURE 4.2
Components of a cell.

4 PRE-LAB EXERCISE 4:
The Cell Cycle _____

During this lab, we will examine the stages of the cell cycle and of mitosis. Use your text and lab atlas to complete the following questions pertaining to the cell cycle and mitosis.

1. Describe the following stages of the cell cycle:

   a. G1:

   b. S:

   c. G2:

   d. M:

2. Describe the events that are occurring in the cell in each of the following phases of mitosis.

STAGE OF MITOSIS	EVENTS OCCURRING IN THE CELL	CELL APPEARANCE
Prophase		
Metaphase		
Anaphase		
Telophase		

EXERCISES

A basic principle that we will revisit several times in our study of anatomy and physiology is:

Form follows function.

This may be stated a variety of ways, and is alternatively called the "principle of complementarity of structure and function." (I personally prefer the simpler version.) This principle refers to the fact that the anatomy (the *form*) is always structured so that it is best suited for the structure's physiology (the *function*).

Imagine if the heart were solid rather than composed of hollow chambers, or if the femur were pencil-thin rather than the thickest bone in the body. They wouldn't be able to carry out their functions of pumping blood and supporting the weight of the body very well, would they? This principle is applicable even on the cellular level, which we will see in this unit.

EXERCISE 1:
Identification of Cell Structures _____

As you learned in the Pre-Lab Exercises, the cell is composed of specialized cellular compartments called *organelles*. Compartmentalization of the cell allows its components to carry out a variety of functions without interfering with one another. In addition to organelles, the cell has structures that help the cell carry out its specific function (remember—form always follows function). These structures were included in the Pre-Lab Exercises and include microvilli, cilia, and flagella.

Note that each individual cell does not contain every organelle or structure. For example, only cells that line certain hollow passages contain cilia, and only cells that are capable of undergoing cell division contain centrioles. In addition, some cell populations contain more of a certain organelle than other populations. The liver contains a high amount of smooth endoplasmic reticulum, and immune cells (phagocytes) house a large number of lysosomes.

Organelles and Cell Structures

Using your textbook and Figures 2.1–2.12 in your lab atlas as a guide, identify the following parts of the cell on models and diagrams. Refer to Pre-Lab Exercise 1 to review the functions of each of these organelles.

1. Cell membrane
   a. Phospholipid bilayer
   b. Integral proteins
   c. Peripheral proteins
   d. Carbohydrates

2. Nucleus
   a. Nuclear membrane
   b. Chromatin and chromosomes
   c. Nucleolus
3. Cytosol
4. Ribosomes
5. Smooth endoplasmic reticulum
6. Rough endoplasmic reticulum
7. Golgi apparatus
8. Lysosomes
9. Peroxisomes
10. Mitochondria
11. Centrioles
12. Microtubules and microfilaments
13. Vesicle
14. Microvilli
15. Cilia
16. Flagella

Model Inventory

Throughout this lab manual, we will use what is termed a *Model Inventory*. In this inventory, you will list the anatomical models or diagrams that you use in lab (if the model is not named, make up a descriptive name) and state which structures you are able to locate on each model. This is particularly helpful for study purposes, as it allows you to return to the proper models to locate specific structures.

MODEL / DIAGRAM	STRUCTURES IDENTIFIED

Building a Cell Membrane

In this exercise, you will visualize the structures of the cell membrane by building a representation of the membrane with modeling clay. Refer to Pre-Lab Exercise 2 for assistance. Use the following color code to build the phospholipid bilayer:

- Phosphate heads: blue
- Fatty acid tails: yellow
- Integral proteins: red
- Peripheral proteins: green
- Carbohydrates: orange

✔ Check Your Understanding

1. Mitochondria contain their own DNA, which encodes 13 proteins. Defects on the mitochondrial DNA can be passed down maternally, leading to a group of diseases called *mitochondrial cytopathies*. Considering the function of mitochondria, explain which cell populations you think would be most affected by such a disease.

2. Another group of diseases affecting a specific organelle are the *lysosomal storage diseases*. What potential problems could be caused by defective lysosomes?

2 EXERCISE 2: Observing Fetal Pig Cheek Cells _____

In this exercise, you will identify some of the above organelles on an actual cell. To see the organelles, you must obtain a sample of cells and then stain them with a dye. In this case, our sample of cells will be obtained from the inside cheek of the fetal pig, and we will stain it with a dye called methylene blue.

Procedure

> ➤ NOTE: SAFETY GLASSES AND GLOVES ARE REQUIRED.

1. Obtain a blank slide and a coverslip.
2. Clean the slide with lens paper.
3. Using an applicator stick, swab the inside cheek of a fetal pig. Do not get large chunks of tissue, as individual cells will not be visible.
4. Wipe the swab with the cheek cells on the blank slide.
5. Place one drop of methylene blue dye onto the slide. (Use gloves—this stuff stains hands readily!) Wait for one minute.
6. Rinse the dye off of the slide with distilled water, and pat dry. The blue dye should be barely visible on the slide. If you see large areas of blue, rinse the slide again or get a new sample of cheek cells.
7. Place a coverslip over the stained area, and use your microscope on high power to find individual cells.
8. Draw an individual cell. Which organelles can you identify?

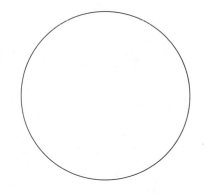

3 EXERCISE 3:
Structure of Different Cell Types _____

The structure of different cell types can vary drastically. Recall from Exercise 1 that cells differ not only in size and shape but also in the types and prevalence of organelles present in the cell. As you look at the following three types of cells, keep in mind that form always follows function.

View prepared slides of red blood cells, sperm cells, and skeletal muscle cells, beginning on low power, and advancing to high power for each slide. Note that sperm cells can be difficult to find, and an oil-immersion lens, if available, is helpful to find the tiny cells. For each slide, draw what you see. Use your text and Figures 2.15–2.20 of your lab atlas for reference.

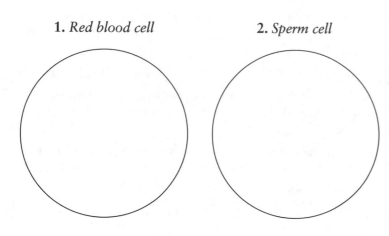

1. *Red blood cell*   2. *Sperm cell*

3. *Skeletal muscle cell*

✔ Check Your Understanding

1. Which cell type lacks a nucleus? What functions would this cell not be able to carry out?

2. Of the three cell types that you examined, the sperm cells were the smallest. Why are these cells so small? What unique cell structure does the sperm cell have? For what purpose?

3. Which type of cell has many nuclei? Why do you think this type of cell needs so many nuclei?

4 EXERCISE 4:
Osmosis and Tonicity _____

In discussing biological solutions, it is useful to describe the concentration of the solution compared to the fluid inside the cell (termed the *intracellular fluid*). This is described as the *tonicity* of the solution. There are three variations of tonicity:

1. *Hypotonic*: This solution has a lower concentration than the intracellular fluid.

2. *Isotonic*: This solution has the same concentration as the intracellular fluid.

3. *Hypertonic*: This solution has a higher concentration than the intracellular fluid.

When cells are placed in solutions that have different tonicities, we are able to watch the process of *osmosis*. Simply defined, osmosis is the movement of solvent (typically water) across a semipermeable membrane from an area of low concentration to an area of high concentration. Be careful not to confuse osmosis with *diffusion*, which is the movement of *solute* from a high concentration to a low concentration. The following experiment will allow you to watch osmosis in action by placing cells—in this case, red blood cells— in solutions of different tonicity.

Procedure

➤ NOTE: SAFETY GLASSES AND GLOVES ARE REQUIRED.

1. Obtain three blank slides, and number them 1 through 3.

2. Using a dropper, place one small drop of animal blood on each slide, taking care to keep your droplet fairly small; otherwise you won't be able to see individual cells.

3. Use a wooden applicator stick to gently spread the droplet around the center of the slide, and place a coverslip on it.

4. On slide 1, place a drop of the 5% dextrose solution on one side of the coverslip. On the other side of the coverslip, hold a piece of lens paper. The lens paper will draw the fluid under the coverslip.

5. Observe the cells under the microscope on high power.

6. Repeat the procedure by placing the 25% NaCl solution on slide 2 and the distilled water on slide 3.

7. Draw and describe what you see with each slide.

Slide 1:

Slide 2:

Slide 3:

Based upon your observations:

1. Which solution was hypotonic? Explain your reasoning.

2. Which solution was isotonic? Explain your reasoning.

3. Which solution was hypertonic? Explain your reasoning.

✔ **Check Your Understanding**

1. Isotonic saline and 5% dextrose in water are solutions that are considered isotonic to human blood. What effect on red blood cells would you expect if a patient were given these fluids intravenously?

2. A solution of 10% dextrose in water is hypertonic to human blood. What would be the consequences if you were to infuse your patient with this solution?

3. Gatorade® and other sports drinks are actually hypotonic solutions. Considering this, explain how these drinks help to rehydrate their users.

5 EXERCISE 5: Mitosis and the Cell Cycle

To maintain the integrity of tissues in the face of continual wear and tear, most cells undergo the process of replacing themselves as they age. This process of cell division is known as *mitosis*. During mitosis the cell divides to produce two identical *daughter cells*. Each daughter cell has the same exact genetic and structural characteristics as the original cell. Mitosis proceeds in four stages, as you learned in the Pre-Lab Exercises. Review these stages prior to reading any further.

Mitosis is part of a larger cycle called the *cell cycle*. During the cell cycle the cell grows (a phase called *G1*), synthesizes DNA (*S phase*), grows again (*G2 phase*), and then, for many cells, enters mitosis (*M phase*). The portions of the cycle from G1–G2, when the cell is not dividing, are collectively called *interphase*. In the Pre-Lab Exercises, you learned the characteristics of the cell during each stage of the cell cycle. Review these characteristics before going further.

Structures of Cell Division

Utilize cell diagrams or models to identify the following structures associated with mitosis. Refer to Exercise 1, the Pre-Lab Exercises, and Figures 2.1, 2.2, 2.22, and 2.23 in your lab atlas for reference.

1. Mitotic spindle
   a. Centrioles
   b. Microtubules (spindle fibers)
2. Nucleus
   a. Nucleolus
   b. Nuclear membrane
3. Chromosomes
   a. Chromatids
   b. Centromere

Cell Models

If your lab has a set of cell cycle models or diagrams, arrange the models in the correct order of the phases of the cell cycle. As an alternative, build a set of cell cycle models with modeling clay, demonstrating the proper order of the cycle. See Figure 2.23 in your lab atlas and the Pre-Lab Exercises for reference.

1. Interphase
2. Mitosis
   a. Prophase
   b. Metaphase
   c. Anaphase
   d. Telophase

Cell Slides

Examine the five phases of the cell cycle using high power on prepared whitefish mitosis slides. Note that every stage of the cell cycle may not be visible on one single slide, so you may have to use more than one slide. Note also that most of the cells you see will be in interphase.

Here, draw and describe what the cell looks like during each phase of the cell cycle, and label your drawing with as many of the structures of cell division (see above) as you can see in each cell. Use your Pre-Lab Exercises and Figure 2.23 in your lab atlas for reference.

1. *Interphase*

Description:

2. *Mitosis*

a. *Prophase*

Description:

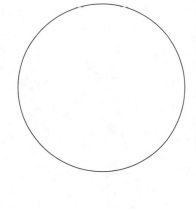

b. *Metaphase*

Description:

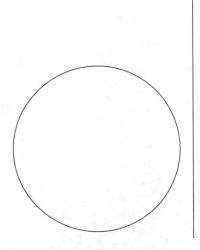

c. *Anaphase*

Description:

d. *Telophase*

Description:

✔ Check Your Understanding

1. What are chromosomes? Why, during mitosis, does the DNA condense into chromosomes?

2. List examples of locations in the body where cell populations undergo rapid mitosis. Why would you expect frequent cell division in these locations?

3. Are there any cell populations that rarely or never undergo mitosis? If yes, give examples. What organelles would these cell populations lack?

5. Histology

■ **OBJECTIVES** _____

Once you have completed this unit, you should be able to:

1. Using the microscope, identify epithelial tissues by number of layers, cell shape, and specializations.

2. Using the microscope, identify and describe a variety of connective tissues.

3. Using the microscope, identify and describe muscle and nervous tissues.

4. Relate tissue structure to tissue function and describe how organs are formed from two or more tissue types.

5. Give examples of organs where each tissue type is found.

■ **MATERIALS** _____

● Tissues slide boxes

● Microscope with three objective lenses

● Lens paper

● Colored pencils

● Five colors of modeling clay

PRE-LAB EXERCISES

Prior to coming to lab, use your text and lab atlas to complete the following exercises.

PRE-LAB EXERCISE 1:
Types of Muscle Tissue

There are three different types of muscle tissue. Use your text to fill in the following chart to determine the differences between each type.

TISSUE TYPE	STRUCTURAL AND FUNCTIONAL CHARACTERISTICS
Cardiac muscle	
Skeletal muscle	
Smooth muscle	

PRE-LAB EXERCISE 2:
Major Tissue Types

Use your text to determine the major types of tissue present in each of the organs listed below. Remember that they are all organs, which means that each of them must have at least two types of tissues.

ORGAN	MAJOR TISSUE TYPES
Urinary bladder	
Blood vessel	
Stomach	
Skin	
Lymph node	
Tendon (and tendon sheath)	
Knee joint and capsule	
Heart	
Trachea	
Esophagus	
Auricle (external ear)	
Brain (and brain coverings)	

EXERCISES

Learning about tissues in general can be highly frustrating. Students typically find that everything looks the same—pink and squiggly. If approached systematically, however, histology is easily understandable and even the most difficult tissues can be deciphered. If you get confused, don't despair. With the help of your lab atlas and a little patience, you can do it!

EXERCISE 1: Epithelial Tissue

Epithelial tissues are those that are found covering body surfaces and lining body passageways and cavities. They are classified in two ways:

1. Number of cell layers:
   a. *Simple epithelium* has only one layer of cells.
   b. *Stratified epithelium* has two or more layers of cells.

2. Shape of cells:
   a. *Squamous* epithelial cells are flat in appearance.
   b. *Cuboidal* epithelial cells are cube-shaped.
   c. *Columnar* epithelial cells are taller than they are long.

Not all epithelial tissues fit this classification scheme. One type of epithelial tissue that you will examine to which this scheme does not apply is called *transitional epithelium*. This type of epithelial tissue is stratified but is not classified by its shape because its cells can change shape. Typically, the cells are dome-shaped, but when the tissue is stretched, the cells flatten out and are squamous in appearance. Another type of epithelial tissue that does not fit neatly into this classification system is *pseudostratified ciliated columnar epithelium*. This epithelium has the appearance of having many layers but actually has only one layer of cells (*pseudo* = false). Note that this type of epithelium has cilia (refer to Unit 4 to review the function of cilia), and the cell shape is always columnar.

As you look at the following epithelial tissues, note a few things that will help you distinguish epithelial tissues from other tissues:

1. Epithelial tissues are all *avascular*. You won't see any blood vessels in epithelium.

2. They are typically on the outer edge of the slide. Keep in mind that most slides have several tissues in each section. To find the epithelial tissue, scroll to one end of the slide or the other.

3. Epithelial tissues consist mostly of cells. We will find that other tissues contain protein fibers and ground substance

surrounding their cells in the extracellular matrix. Epithelial tissue contains very little extracellular matrix and instead has mostly cells that are packed together tightly. If you aren't sure if something is a cell, look for a nucleus. If you can see a nucleus, you typically will also be able to see the cell membrane surrounding the cell, which will help you define borders between cells.

Epithelial Tissue Slides

For the following types of epithelial tissue, examine prepared slides and use colored pencils to draw what you see, then (a) describe what you see and (b) give examples of organs where the tissue is found. For study purposes, it is also helpful to write the name of the slide that contains the tissue in question (e.g., the slide for stratified squamous, keratinized epithelium may be called "palmar skin"). Use your text and Figures 3.1–3.8 in your atlas for reference.

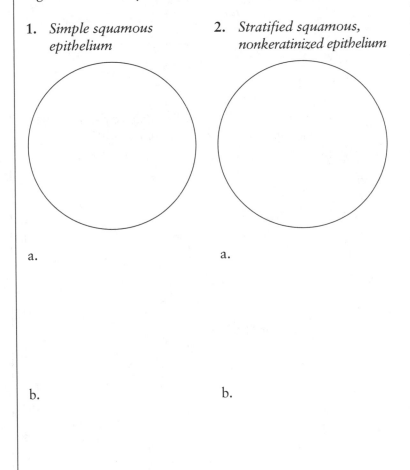

1. *Simple squamous epithelium*

2. *Stratified squamous, nonkeratinized epithelium*

a.

a.

b.

b.

3. Stratified squamous, keratinized epithelium

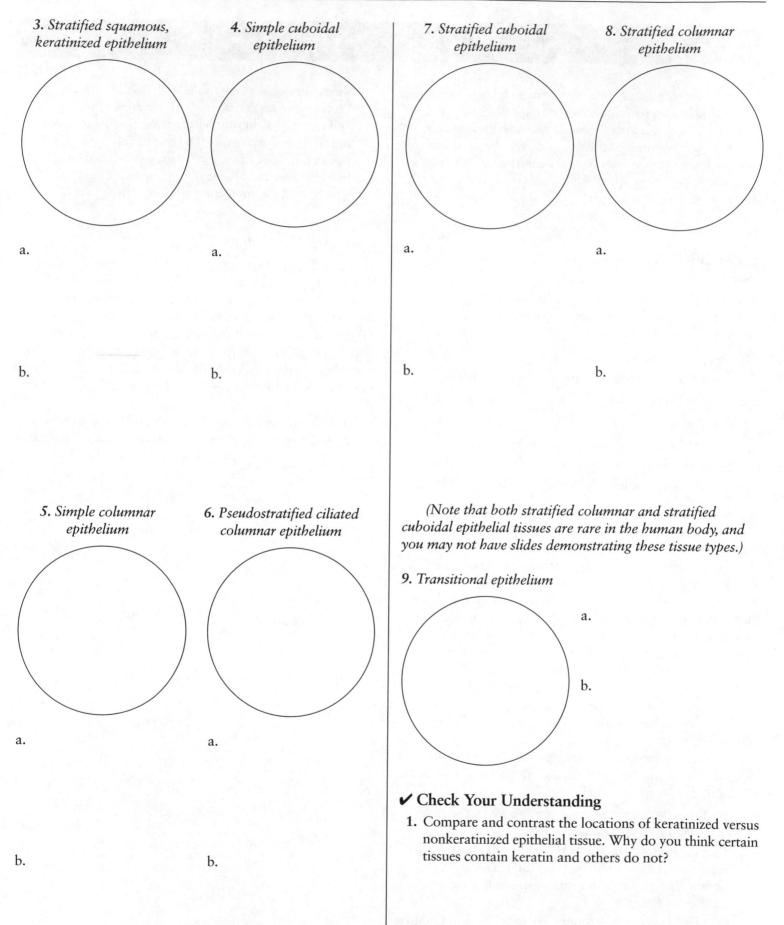

a.

b.

4. Simple cuboidal epithelium

a.

b.

7. Stratified cuboidal epithelium

a.

b.

8. Stratified columnar epithelium

a.

b.

5. Simple columnar epithelium

a.

b.

6. Pseudostratified ciliated columnar epithelium

a.

b.

(Note that both stratified columnar and stratified cuboidal epithelial tissues are rare in the human body, and you may not have slides demonstrating these tissue types.)

9. Transitional epithelium

a.

b.

✔ **Check Your Understanding**

1. Compare and contrast the locations of keratinized versus nonkeratinized epithelial tissue. Why do you think certain tissues contain keratin and others do not?

2. Compare and contrast the locations of simple versus stratified epithelial tissues. Why do you think certain tissues are simple and others are stratified? (*Hint*: Form follows function.)

2 EXERCISE 2: Connective Tissue _____

Connective tissues are found throughout the body. They have a variety of functions, most serving to *connect*, as their name implies (blood is an exception). All connective tissues stem from a common embryonic tissue called *mesenchyme*. The four general types of connective tissue (CT) we will examine are:

1. *Connective tissue proper*: This class includes loose (areolar) CT, reticular CT, adipose tissue, elastic tissue, dense regular CT, and dense irregular CT.

2. *Cartilage*: This class includes hyaline cartilage, fibrocartilage, and elastic cartilage.

3. *Bone*

4. *Blood*

The following points will help you to distinguish connective tissues from other tissues:

1. The cells typically are not densely packed together. You will see a large amount of space between connective tissue cells. Remember to look for the nucleus and cell membrane to discern the borders of the cell.

2. Generally, connective tissue has large amounts of *ground substance* and *protein fibers* in the extracellular matrix, particularly connective tissue proper. One exception is *adipose tissue*, which consists of densely packed *adipocytes*, which are filled with a large lipid droplet.

3. Many types of CT can be distinguished by the types of fibers that they contain:
   a. *Reticular fibers*: These are thin fibers that typically stain black. They create networks to support small structures such as vessels and nerves. Look for reticular fibers in loose and reticular CT.
   b. *Collagen fibers*: These are thick, tough fibers composed of the protein collagen. Depending on the stain used, they are often stained pink. The tough nature of collagen fibers suits them well for supportive functions. Look for collagen fibers in fibrocartilage, dense regular, and dense irregular CT. They also are found in bone tissue; however, bone is rarely stained so that collagen fibers are visible.

   c. *Elastic fibers*: These fibers are also fairly thick, but they have the notable property of being elastic (bet you couldn't have guessed that!). This allows tissues to reassume their original shape when deformed. They often stain purple and are found in elastic tissue and elastic cartilage.

For each of the following types of connective tissue, examine a prepared slide of the tissue type, draw what you see using colored pencils, and then (a) describe what you see and (b) give examples of organs where the tissue is found. Use your text and Figures 3.9–3.33 in your atlas as a guide.

Connective Tissue Proper

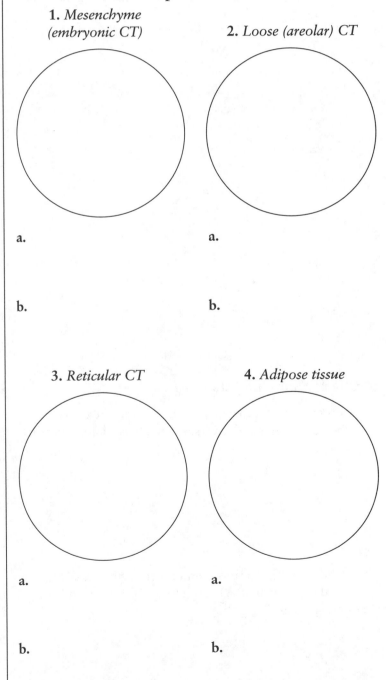

1. Mesenchyme (embryonic CT)

2. Loose (areolar) CT

a.

a.

b.

b.

3. Reticular CT

4. Adipose tissue

a.

a.

b.

b.

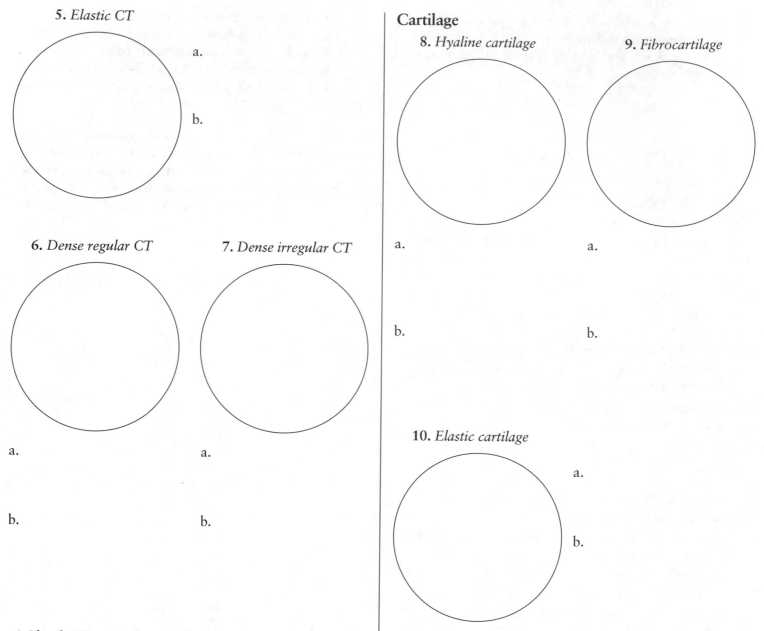

5. *Elastic CT*

a.

b.

6. *Dense regular CT*

a.

b.

7. *Dense irregular CT*

a.

b.

Cartilage

8. *Hyaline cartilage*

a.

b.

9. *Fibrocartilage*

a.

b.

10. *Elastic cartilage*

a.

b.

✔ **Check Your Understanding**

1. Compare the arrangement of the extracellular protein fibers in dense regular versus dense irregular connective tissues.

2. Based upon the locations and functions of dense regular and dense irregular CT, explain the difference in fiber arrangement.

✔ **Check Your Understanding**

Something you should notice when comparing the three types of cartilage is that they vary according to the number and types of extracellular fibers.

1. Which type of cartilage has the fewest fibers and is made primarily of ground substance? Why do you think this is so?

2. The formation of fibrocartilage is a common response to injury of hyaline cartilage. Do you think that fibrocartilage would provide an articular surface (i.e., the cartilage in joints) that is as smooth as the original hyaline cartilage? Why or why not?

2. Note on the slide of bone tissue the densely packed rings of bone matrix. Considering the primary functions of bone, explain how its form follows its function.

Bone and Blood

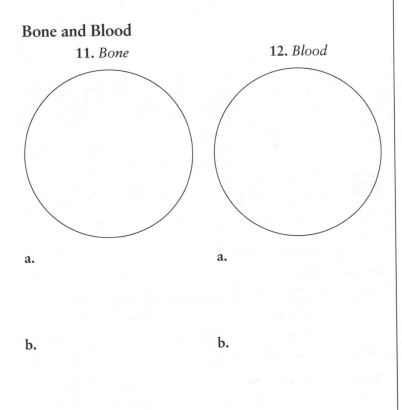

11. *Bone*

12. *Blood*

a.

a.

b.

b.

✔ Check Your Understanding

1. A tissue is composed of two or more similar cell types functioning together. In the case of blood, what cell types do you see? What are their functions?

3 EXERCISE 3: Muscle Tissue_____

Here are some hints to help you identify the three types of muscle:

1. If the cells are long and *striated*, it is *skeletal muscle*. Skeletal muscle cells are some of the longest cells in the body, with one cell extending the entire length of the muscle.

2. If the cells are short, fat, and striated, and adjacent cells are held together by *intercalated discs* (see Figure 3.37 in your atlas), it is *cardiac muscle*.

3. If the cells are spindle-shaped with one nucleus in the center of the spindle, with no noticeable striations, it is *smooth muscle*. Sometimes students say that smooth muscle somewhat resembles squamous cells.

For each of the following tissues, examine a prepared slide of the tissue and use colored pencils to draw what you see, then (a) describe what you see and (b) give examples of organs where the tissue is found. Refer to the table you filled out in Pre-Lab Exercise 1 and Figures 3.34–3.39 in your atlas as a guide.

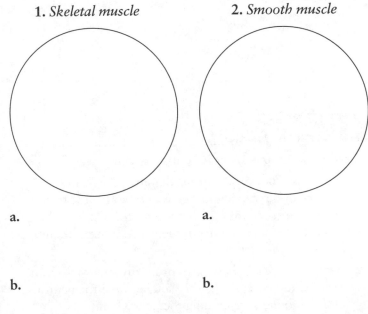

1. *Skeletal muscle*

2. *Smooth muscle*

a.

a.

b.

b.

3. *Cardiac muscle*

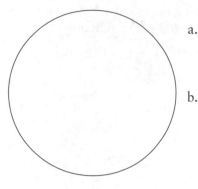

a.

b.

✔ Check Your Understanding

1. You may have noticed that skeletal muscle cells have multiple nuclei. Considering their size, explain why these cells have so many nuclei.

2. Both cardiac and smooth muscle cells have only one nucleus. Why do you think that this is so?

4 EXERCISE 4: Nervous Tissue _____

Nervous tissue consists of two primary cell types:

1. *Neurons.* These are the cells that are responsible for sending and receiving messages. On your slide they are the larger of the two cell types. You will note long projections extending from the neuron; these are termed *axons* and *dendrites.*

2. *Neuroglial cells.* These have primarily supportive functions. They are considerably smaller than the neurons and are found surrounding the neuron and its processes.

Observe the slide with nervous tissue (may be called "motor neuron smear") and use colored pencils to draw what you see, then (a) describe what you see and (b) give examples of locations in the body where the tissue is found. Use your textbook and Figures 3.41 and 3.49–3.51 in your atlas for reference.

1. *Nervous tissue*

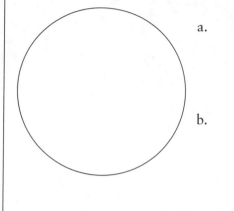

a.

b.

5 EXERCISE 5: Building an Organ _____

We learned that organs consist of two or more tissue types, and now we will visualize this firsthand. For this exercise, your lab group will build one or more organs out of modeling clay, using this color code:

Yellow = epithelial tissue (Be sure to use more than one layer if the tissue is stratified!)

Blue   = connective tissue proper

Orange = cartilage

Red   = muscle tissue

Green   = nervous tissue (Don't forget that nearly every organ in the body has nerves supplying it. To place the nerves properly, find out which tissue layer of the organ is innervated.)

Organs from which to choose:

- Heart
- Blood vessel
- Trachea
- Urinary bladder
- Esophagus
- Uterus

6. Integumentary System

OBJECTIVES

Once you have completed this unit, you should be able to:

1. Identify anatomical structures of the integumentary system on models and diagrams.

2. Identify histological structures of the skin and hair on prepared microscope slides.

3. Map the distribution of touch receptors on different areas of the body.

4. Perform and interpret fingerprinting procedures.

■ **MATERIALS**

- Anatomical models: skin sections
- Microscope slides: thick skin and thin skin
- Colored pencils
- Water-soluble marking pens
- Rulers
- Blank glass slides
- Fingerprinting supplies:
    - Inkpads
    - Fingerprint cards
    - Dusting powder
    - Dusting brush
    - Fingerprint lifting tape

PRE-LAB EXERCISES

Prior to coming to lab, complete the following exercises using Chapter 4 in your lab atlas and your text for reference.

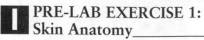**PRE-LAB EXERCISE 1:**
Skin Anatomy

Label Figure 6.1 with the terms from Exercise 1, and use colored pencils to color-code the structures. See Figure 4.1 in your lab atlas for reference.

FIGURE 6.1
Diagram of skin section.

2 PRE-LAB EXERCISE 2:
Hair and Nail Anatomy

Label and color-code the following diagrams of a hair and a nail with the terms from Exercise 1. See Figures 4.6, 4.10, and 4.11 in your lab atlas for reference.

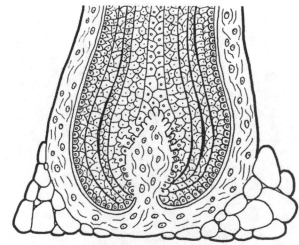

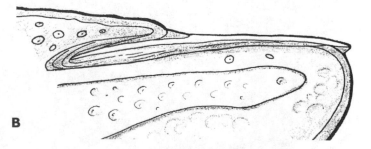

A

B

FIGURE 6.2
(A) Diagram of hair; (B) diagram of nail.

3 PRE-LAB EXERCISE 3:
Functions of Skin Structures

Describe the function of each skin structure listed in the chart below.

SKIN STRUCTURE	FUNCTION
Epidermis	
Epidermal layers:	
1. Stratum corneum	
2. Stratum lucidum	
3. Stratum granulosum	
4. Stratum spinosum	
5. Stratum basale	
Epidermal cells:	
1. Keratinocytes	
2. Melanocytes	
3. Merkel cells (discs)	
Dermis	
Dermal layers:	
1. Papillary layer	
2. Reticular layer	
Dermal structures:	
1. Dermal papillae	
2. Pacinian (lamellated) corpuscle	
3. Meissner's (tactile) corpuscle	
Hypodermis and subcutaneous fat	
Accessory structures:	
1. Hair	
a. Hair follicle	
b. Hair shaft	
c. Arrector pili muscle	
2. Sweat glands	
3. Sebaceous glands	
4. Nails	

EXERCISES

The Integumentary System

The integumentary system is composed of the skin (the *integument*) and its accessory organs: the hair, glands, and nails. The skin, the largest organ in the body, is composed of two layers, the *epidermis* and the *dermis*, each of which may be further subdivided into thinner layers. The tissue beneath the dermis, sometimes called the *hypodermis* or the subcutaneous tissue, connects the skin to the underlying tissues and is not technically considered to be part of the integument. In the following exercises you will examine the skin and accessory structures, and apply your knowledge of the skin to perform fingerprinting procedures.

EXERCISE 1:
Skin Anatomy and Accessory Structures _____

The epidermis contains layers (or *strata*) of stratified squamous keratinized epithelium. From superficial to deep, the layers are:

1. *Stratum corneum*: This layer is composed of dead cells called *keratinocytes*. On microscopic examination, the cells of the stratum corneum bear little resemblance to living cells and instead have a dry, flaky appearance. These cells contain a hard protein called *keratin*, which provides protection for the underlying layers.

2. *Stratum lucidum*: This is a single layer of translucent, dead cells that is found only in the skin of the palms and the soles.

3. *Stratum granulosum*: The superficial cells of the stratum granulosum are dead, but the deeper cells are alive. This layer is named for the cells' cytoplasmic granules, which contain the protein keratin and an oily, acidic, waterproofing substance.

4. *Stratum spinosum*: The first actively metabolizing cells are encountered in the stratum spinosum. The pigment *melanin* is found in this layer, which provides protection from UV light and also tempers production of vitamin D so the body does not overproduce it.

5. *Stratum basale*: The deepest layer, the stratum basale, contains a single layer of actively dividing cells. When grouped with the stratum spinosum, it often is called the *stratum germinativum* because these two layers are the only ones undergoing mitosis.

Why does the epidermis have so many dead cells? Recall that the epidermis is composed of epithelial tissue, and that epithelial tissue is avascular (it has no blood supply). All epithelial tissues require oxygen and nutrients to be delivered from the deeper tissues. In the case of the epidermis, this deeper tissue is the dermis. Only the cells of the deeper parts of the stratum granulosum, the stratum spinosum, and the stratum basale are close enough to the blood supply in the dermis to get adequate oxygen and nutrients for survival. As the cells migrate farther away from the blood supply, they begin to die.

Immediately deep to the stratum basale of the epidermis is the dermis, which is composed of highly vascular connective tissue, and contains two layers:

1. *Papillary layer*: The superficial papillary layer is composed of loose areolar connective tissue. It contains fingerlike projections called the *dermal papillae* that project into the epidermis. These dermal papillae contain touch receptors called *Meissner's corpuscles*, as well as capillary loops that provide blood supply to the avascular epidermis.

2. *Reticular layer*: The thick reticular layer is composed of dense irregular connective tissue proper. It houses structures such as oil-producing *sebaceous glands*, *sweat glands*, blood vessels, and deep pressure receptors such as *Pacinian corpuscles*.

The integument contains numerous accessory structures, including hair, nails, sweat glands, and sebaceous glands. Hair and nails both consist of dead, keratinized epidermal cells and are embedded in the dermis. Sebaceous glands and sweat glands are exocrine glands that secrete their products onto the skin's surface. Sebaceous glands secrete sebum (oil) into a hair follicle, and sweat glands secrete sweat through a small pore.

Identify the following structures of the integumentary system on models and diagrams using the Pre-Lab Exercises and Chapter 4 in your lab atlas as a guide.

Skin Structures

1. Epidermal layers
   a. Stratum corneum
   b. Stratum lucidum
   c. Stratum granulosum
   d. Stratum spinosum
   e. Stratum basale
2. Dermal layers
   a. Papillary layer
   b. Reticular layer
3. Dermal papillae
4. Blood vessels

5. Nerves:
   a. Pacinian (lamellated) corpuscle
   b. Meissner (tactile) corpuscle
   c. Merkel discs
6. Hypodermis
   a. Subcutaneous fat
   b. Subcutaneous blood vessels

Accessory Structures

1. Sweat gland
2. Hair
   a. Hair follicle
   b. Hair shaft
   c. Arrector pili muscle
3. Nail
   a. Eponychium
   b. Nail matrix
   c. Nail fold
   d. Nail plate
   e. Lunula
4. Sebaceous gland

Model Inventory

As you view the anatomical models and diagrams in lab, list them on the inventory below, and state which of the above structures you are able to locate on each model.

MODEL/DIAGRAM	STRUCTURES IDENTIFIED

✔ Check Your Understanding

1. Explain why a superficial skin scrape (such as a paper cut) doesn't bleed. Why don't you bleed when a hair is pulled out?

2. Shampoos and hair conditioners often claim to have nutrients and vitamins that your hair needs in order to grow and be healthy. Taking into account the composition of hair, do you think these vitamins and nutrients will be beneficial? Why or why not?

2 EXERCISE 2:
Histology of Integument _____

In this exercise we will examine prepared slides of skin from different regions of the body in order to compare and contrast two types of skin: (1) thick skin, found on the palms of the hands and soles of the feet, and (2) thin skin, found virtually everywhere else.

Before moving on to the next part, you may wish to review the basics of microscopy from Unit 3. Remember to follow a step-by-step approach when examining the slides: look at the slide with the naked eye first, and then begin your examination on low power, advancing to higher power to see more details.

Thick Skin

Obtain a prepared slide of thick skin (which may be labeled "Palmar Skin"), and, after examining it with the naked eye to get oriented, place it on the stage of the microscope. First, scan the slide on low power. You should be able to see the epidermis, with its superficial layers of dead cells, and the dermis, with its clusters of collagen bundles that make up the dense irregular connective tissue proper. Advance to higher power to see the cells and associated structures in greater detail. Following are details that you should note that will make it easier to discern thick skin from thin skin:

● The epidermis of thick skin is notably thicker, in particular the stratum corneum.

● You may be able to find the stratum lucidum, which is a thin layer of clear cells between the stratum corneum and the stratum granulosum.

● Thick skin has no hair follicles, arrector pili muscles, or sebaceous glands. (This is always the dead giveaway.)

Use your colored pencils to draw what you see in the field of view (you will be able to see the most structures on low power). Label your drawing with the terms on page 45, using Figure 4.5 in your atlas for reference.

Terms

1. Epidermis
   a. Stratum corneum
   b. Stratum lucidum
   c. Stratum granulosum
   d. Stratum spinosum
   e. Stratum basale
2. Dermis
   a. Dermal papillae
   b. Collagen bundles

Thin Skin

Obtain a prepared slide of thin skin (which may be called "Scalp Skin"). As before, examine the slide with the naked eye, and then scan the slide on low power, advancing to higher power as needed to see the structures more clearly. Following are some features of thin skin that will help you discern it from thick skin:

- Each layer of the epidermis overall is much thinner, especially the stratum corneum, which appears as flaky cells that are sloughing off.

- Thin skin lacks a stratum lucidum.

- You should be able to find features that are absent in thick skin, including abundant hair follicles and sebaceous glands. Note that the hair follicle is composed of epidermal tissue that is continuous with the stratum basale, and dives deep into the dermis.

- You also may see arrector pili muscles, which appear as thin bands of smooth muscle.

Use your colored pencils to draw what you see in the field of view (you will be able to see the most structures on low power). Label your drawing with the preceding terms, using Figures 4.7 and 4.10 in your atlas for reference.

Terms

1. Epidermis
   a. Stratum corneum
   b. Stratum granulosum
   c. Stratum spinosum
   d. Stratum basale
2. Dermis
   a. Dermal papillae
   b. Collagen bundles
3. Hair follicle
4. Sebaceous gland
5. Sweat gland
6. Arrector pili muscle

3 EXERCISE 3:
Touch Receptor Distribution _____

In the Pre-Lab Exercises, you described the function of fine-touch receptors in the epidermis and the dermis called Meissner's corpuscles and Merkel discs. It is possible to map the distribution of these receptors using an instrument called a Semmes–Weinstein monofilament, which applies 10 grams of force as it is pressed on the skin. Ten grams of force is generally accepted as the maximum amount of force required to activate the Meissner's corpuscles and Merkel discs. If a monofilament is not available, a similar instrument may be made using a stiff-bristle hair glued to a toothpick.

Procedure

1. Use a centimeter ruler to measure out a 2-cm square on your partner's anterior forearm. Mark this square with a water-soluble marking pen.

2. Have your partner close his or her eyes.

3. Apply the Semmes–Weinstein monofilament to one of the corners of the square. Press only until the monofilament bends slightly.

4. If your partner can perceive the sensation of the monofilament, mark this spot with a different color of water-soluble marker.

5. Repeat this process for the remainder of the square, advancing the monofilament 1–2 millimeters each time. Be certain to apply the same amount of pressure with each application.

6. When you have completed the square, record the number of receptors that were found on the anterior forearm.

   Number of receptors on the anterior forearm: _____

7. Repeat this process on the posterior shoulder. Record the number of receptors found on this location.

   Number of receptors on the posterior shoulder: _____

✔ Check Your Understanding

1. Would you have expected to have found more receptors on the anterior forearm or on the posterior shoulder? Why?

2. In which areas of the body would you expect to find the most Merkel discs and Meissner's corpuscles?

4 EXERCISE 4:
Fingerprinting _____

Fingerprinting was not a widely accepted means of identification in the United States until 1902, when the New York Civil Service began collecting and using fingerprints. Prior to this, fingerprints had been widely studied and were even used to solve a murder case in Argentina in 1892.

Fingerprints can be broadly classified into ridge flow patterns: loop, arch, and whorl (see Figure 6.3A). Once the ridge flow pattern has been identified, the characteristics of the ridges, also called *Galton points*, are examined. The three basic ridge characteristics are the ridge ending, the island (or the dot), and the bifurcation (see Figure 6.3B). In professional laboratories, the fingerprint is further analyzed on the basis of minute details, or "points," in the fingerprint. Much of this is currently done with the help of computers.

In this exercise you will be playing the role of detective, searching for a "thief" among your group members (which may end up being you!). Your group will determine the identity of the thief by analyzing the ridge flow patterns and ridge characteristics of the fingerprints of your group members.

Procedure

1. Assemble into groups of a minimum of four people.

2. Obtain one blank glass slide, and, handling it by the edges, clean it off with a paper towel and water.

3. Once the slide is clean and dry, place two or three large fingerprints (use different fingers, preferably on each hand) somewhere on one side of the slide.

4. After placing one or more fingerprints on the slide, put on gloves and use the sharpie at your table to write your initials on the upper righthand corner of the slide.

5. Bring the slides for the whole table to your instructor, and your instructor will pick one slide from the group and designate that person as the "thief."

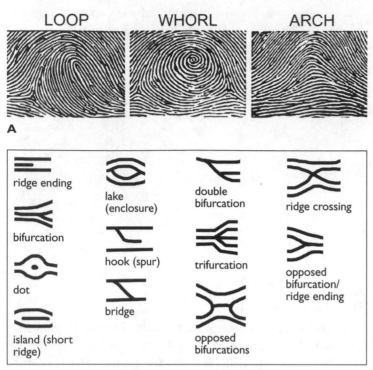

A

B

FIGURE 6.3
(A) Diagrams of the three basic patterns of fingerprints, and
(B) the Galton points.

6. The thief is one of your group members, so you are all suspects. Each member of the group must be fingerprinted (using the ink and cards provided):

   a. Obtain a blank fingerprinting card or a blank sheet of paper.

   b. Roll each finger individually on the inkpad, rolling from right to left. Be certain not to get excessive ink on your fingers.

   c. Roll each finger individually on the fingerprinting card, rolling slowly from left to right.

7. When all of the fingerprinting is completed, don a pair of latex gloves.

8. Use a dusting brush to lightly sweep some dusting powder onto the slide containing the thief's fingerprints. Use the dusting powder *sparingly*, as too much powder will obscure your prints. Brush away the excess dusting powder. (You can tap the slide on its side to help with this.)

9. Place a piece of fingerprint lifting tape on the fingerprints that have appeared. Pat the tape into place firmly, and slowly remove it. Place the tape on a blank sheet of white paper for comparison.

10. With the help of Figures 6.3A and B, identify the thief by comparing the fingerprints lifted from the slide with the fingerprints taken from the suspects. Be careful not to

tell your group mates which fingers you used. Remember —you don't want to be caught!

Who is the thief from your table?

How did you arrive at this conclusion?

✔ Check Your Understanding

1. Which anatomical feature creates the epidermal ridges that allow fingerprinting? What purpose does this feature serve?

2. Why do you have unique markings on your fingers, toes, palms, and soles but not on any other part of the body?

3. Why do you leave behind fingerprints when you touch certain surfaces?

7. Introduction to the Skeletal System

■ **OBJECTIVES**

Once you have completed this unit, you should be able to:

1. Identify the structures and components of osseous tissue on models and slides.

2. Explain the role of minerals and protein fibers in the function of osseous tissue.

3. Classify bones according to their shape.

4. Identify the parts of a long bone.

5. Define and provide examples of various bone markings.

■ **MATERIALS**

● Models of the osteon

● Slides of cortical and cancellous bone

● Small samples of fresh bone

● Nitric acid

● Disarticulated bones

● Fresh long bone, sectioned

● Articulated skeleton

● Colored pencils

PRE-LAB EXERCISES

Prior to coming to lab, complete the following exercises using Chapter 5 in your lab atlas and your text for reference.

1 PRE-LAB EXERCISE 1: Structures of the Osteon _____

Figure 7.1 is a diagram of an osteon, the functional unit of cortical bone. Label the structures of the osteon with the terms from Exercise 1, and use colored pencils to color-code the structures (note: you may wish to wait until your lab period in order to color-code the diagram in a similar manner to the models in lab). See Figure 3.29 in your atlas for reference.

2 PRE-LAB EXERCISE 2: Functions of the Osteon Structures_____

Now that you have labeled a diagram of an osteon, use your text to learn the function of each part you labeled. List the function in the table below.

3 PRE-LAB EXERCISE 3: Definitions of the Bone Markings _____

On page 51 is a chart containing common bone markings. Define each marking, using your text and Table 6.1 in your lab atlas for reference.

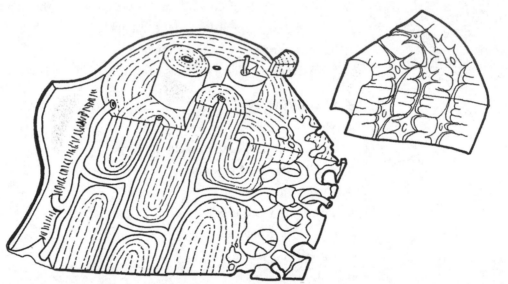

FIGURE 7.1
Diagram of cortical bone.

OSTEON STRUCTURE	DESCRIPTION	FUNCTION
Lamellae		
Haversian (central) canal		
Volkmann's (perforating) canal		
Lacunae		
Osteocytes		
Canaliculi		
Periosteum		
Endosteum		

BONE MARKING	DEFINITION
Alveolus	
Condyle	
Crest	
Epicondyle	
Facet	
Foramen	
Fissure	
Fossa	
Fovea	
Head	
Meatus	
Process	
Sinus	
Spine	
Sulcus	
Trochanter	
Tuberosity	
Tubercle	

EXERCISES

The *skeletal system* consists of the bones and their cartilages. Each bone, which contains osseous tissue and also nervous, epithelial, dense connective, and adipose tissue, is considered an organ. In the following exercises you will be introduced to the skeletal system, including the histology of osseous tissue, the various shapes of bone, the structures associated with long bones, and the markings and processes that are found on bones.

EXERCISE 1:
Anatomy of Osseous Tissue _____

There are two general types of bone tissue, called *cortical bone* and *cancellous bone* (also known as *compact* and *spongy* bone, respectively). The hard, dense cortical bone found on the outer region of the bone is composed of repeating units called *osteons*. It serves as the structural support of the bone. Cancellous bone, found on the inner region of the bone, has a latticework-type structure that does not contain osteons but instead has tiny bone spicules called *trabeculae*. The latticework structure of cancellous bone allows it to house another important tissue, the *bone marrow*.

Following is a list of cortical bone structures you will identify in lab:

Cortical Bone Structures

1. Osteon (Haversian system)
2. Central (Haversian) canal
3. Perforating (Volkmann's) canal
4. Lamellae
5. Lacunae
   a. Osteocytes
6. Canaliculi
7. Periosteum
   a. Sharpey's fibers
8. Endosteum

Cancellous Bone Structures:

1. Trabeculae (bone spicules)
2. Marrow
3. Endosteum
4. Lacunae
5. Diplöe
6. Osteocytes

Model Inventory

As you view the anatomical models in lab, list them on the inventory below, and state which of the above structures you are able to locate on each model.

MODEL	BONE STRUCTURES IDENTIFIED

✔ Check Your Understanding

1. Considering the primary function of cortical bone, explain its structure.

2. Considering the primary function of cancellous bone, explain its structure.

2 EXERCISE 2: Histology of Osseous Tissue _____

View a prepared slide of cortical bone, first on medium, then on high power. The structures on the slide should look similar to the osteon models you viewed in lab already.

Use colored pencils to draw a picture of what you see under the microscope, and label your drawing with the cortical bone structures from Exercise 1. Use your text and Figures 3.27, 3.28, 3.29, and 3.31 from your lab atlas for reference.

3 EXERCISE 3: Chemical Components of Bone Tissue _____

Bone, a type of connective tissue, is composed of two primary chemical components:

1. The *organic component* consists mostly of protein fibers such as *collagen* (see Figure 3.28 in your lab atlas). Collagen provides a supportive network that gives the bone tensile strength (the ability to withstand stretching forces).

2. *The inorganic component* consists mostly of calcium in the form of calcium hydroxyapatite crystals. The inorganic component provides the bone with compressional strength (the ability to withstand compressive forces).

In this exercise you will compare the effects of removing different chemical components of bone. One sample has been heated to destroy the organic components of the bone, and another has been treated with nitric acid, which dissolves calcium crystal, destroying the inorganic component of bone.

Procedure

➤ NOTE: SAFETY GLASSES AND GLOVES ARE REQUIRED.

1. Obtain three pieces of bone:
   a. One piece of bone that has been baked in an oven for a minimum of two hours,

b. One piece of bone that has been soaked in nitric acid, and

c. One piece of untreated bone.

2. Place the three pieces of bone side by side. Do they look different in appearance from one another?

3. Compress or squeeze each sample using your fingers. What happens to:

a. The heated bone?

b. The bone treated with nitric acid?

c. The untreated bone?

✔ **Check Your Understanding**

1. *Osteogenesis imperfecta* is a congenital condition in which collagen synthesis is defective. Of the three samples of bone above, which is most similar to the bones in osteogenesis imperfecta? What symptoms would you expect to find in this disease?

2. The diseases *rickets* and *osteomalacia* result from insufficient vitamin D intake, which decreases the amount of calcium available for synthesis of the inorganic component of bone. Of the three samples of bone above, which is the most similar to the bones in rickets and osteomalacia? What symptoms would you expect to find in these diseases?

**4 EXERCISE 4:
Classification of Bones by Shape** _____

One way in which bones are classified is by their shape. Using your text and Figure 5.1 in your lab atlas as a reference, obtain several samples of disarticulated bones and classify them according to this scheme. Identify examples of:

1. Long bones
2. Short bones
3. Irregular bones
4. Flat bones

SHAPE	EXAMPLES
Long bones	
Short bones	
Irregular bones	
Flat bones	

**5 EXERCISE 5:
Structure of Long Bones**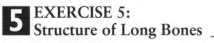

One of the classes of bones discussed above, long bones, has parts that are unique to its class. Use your text and Figure 6.1 in your lab atlas to identify the following structures of long bones on fresh and sectioned specimens and X-rays (if available). Label the parts of a long bone on Figure 7.2.

1. Diaphysis
2. Epiphysis
3. Medullary canal
4. Epiphyseal plate (may be visible only on X-ray)
5. Epiphyseal line
6. Cortical (compact) bone
7. Cancellous (spongy) bone

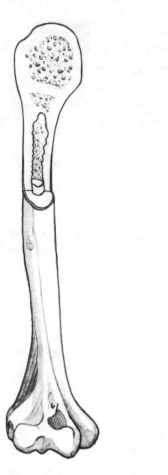

FIGURE 7.2
Diagram of a
long bone.

✔ Check Your Understanding

Often, in children, an open epiphyseal plate, which indicates active growth, is misread on X-rays as a fracture.

1. Of which tissue type is the epiphyseal plate composed?

2. Does this tissue appear the same as bone on an X-ray? (*Hint*: Think about an X-ray of the knee. Can you see the articular cartilage?)

3. Considering this, explain why open epiphyseal plates are often misread on X-rays as fractures.

6 EXERCISE 6: Bone Markings

In the Pre-Lab Exercises, you learned about the different markings, processes, and indentations found in the bones. To familiarize yourself with each of the bone markings, locate an example of each and record these in the chart below, using your lab atlas and a set of disarticulated bones.

BONE MARKING	EXAMPLE OF BONE	NAME OF BONE MARKING
Alveolus		
Condyle		
Crest		
Epicondyle		
Facet		
Fissure		
Foramen		
Fossa		
Fovea		
Head		
Meatus		
Process		
Sinus		
Spine		
Sulcus		
Trochanter		
Tuberosity		
Tubercle		

8. Skeletal System

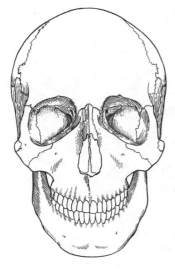

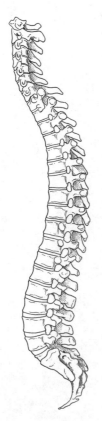

■ **OBJECTIVES**

Once you have completed this unit, you should be able to:

1. Identify bones and markings of the axial skeleton.

2. Identify bones and markings of the appendicular skeleton.

3. Build a skeleton using disarticulated bones.

■ **MATERIALS**

- Skeletons, articulated
- Skeletons, disarticulated
- Vertebral column
- Skulls
- Male and female pelvises
- Colored pencils

PRE-LAB EXERCISES

The skeletal system has two divisions: the axial skeleton and the appendicular skeleton. The *axial skeleton* is composed of the bones of the head, neck, and trunk—specifically the cranial bones, the facial bones, the vertebral column, the hyoid bone, the sternum, and the ribs. The *appendicular skeleton* consists of the bones of the arms, the legs, the pectoral girdle (the bones forming the shoulder joint), and the pelvic girdle (the bones forming the pelvis and hip joint).

Use your text and lab atlas to label the following figures depicting the axial and appendicular skeletons.

▐ PRE-LAB EXERCISE 1: Bones of the Skull

Figure 8.1 shows several views of the skull. Use your colored pencils to color-code and label each of the major bones of the skull, as well as the bone markings indicated in Exercise 1 of this unit. To avoid confusion, be certain to be consistent with your colors when moving from one diagram to the next! Use Figures 5.4–5.25 in your lab atlas for reference.

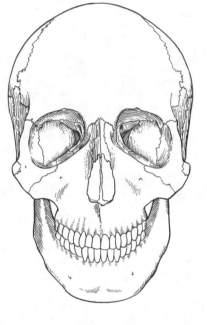

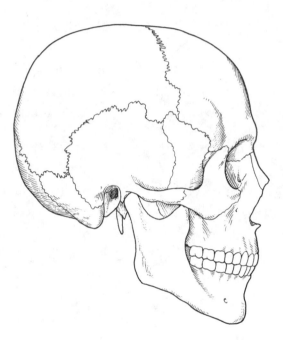

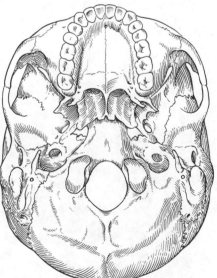

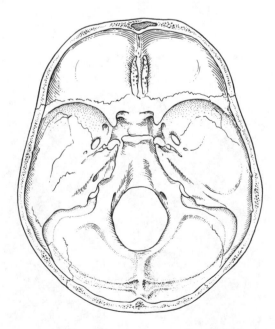

FIGURE 8.1
Anterior, lateral, inferior, and internal views of the skull.

2 PRE-LAB EXERCISE 2:
Remaining Bones of the Axial Skeleton _____

Figure 8.2 contains diagrams of the remaining bones of the axial skeleton. Color-code and label these in a manner similar to the bones of the skull with the bones and bone markings indicated in Exercise 2. Refer to Figures 5.26–5.37 in your lab atlas for assistance.

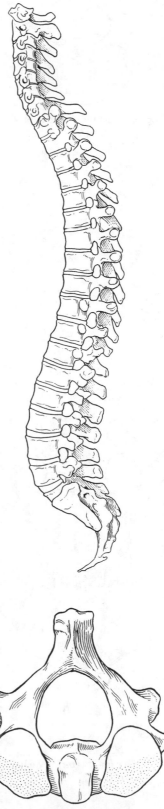

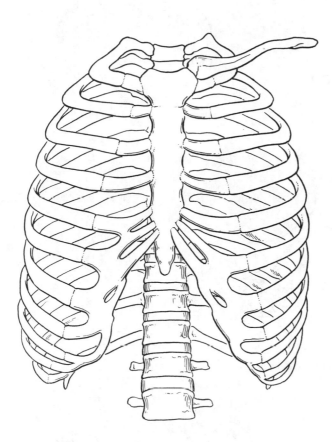

Sternum and thoracic cage

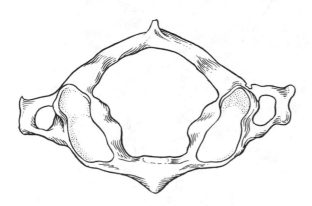

Atlas—Superior view

Axis—Superior view

FIGURE 8.2
Sternum and thoracic cage, and bones of the vertebral column.

(Continued)

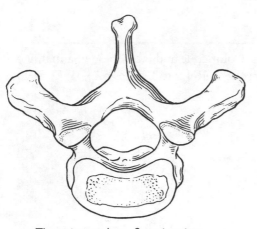

Thoracic vertebra—Superior view

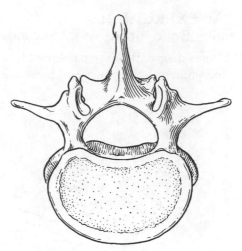

Lumbar vertebra—Superior view

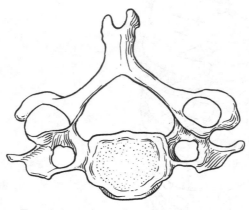

Typical cervical vertebra—Superior view

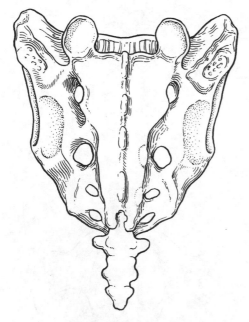

Sacrum—Anterior view

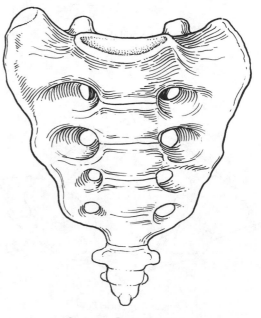

Sacrum—Posterior view

FIGURE 8.2 *Continued.*

3 PRE-LAB EXERCISE 3:
Appendicular Skeleton

Figure 8.3 shows anterior and posterior views of the appendicular skeleton. Color-code and label this figure in a manner similar to Pre-Lab Exercises 1 and 2, using the structures of Exercise 3. See Figures 6.3–6.26 in your lab atlas for reference.

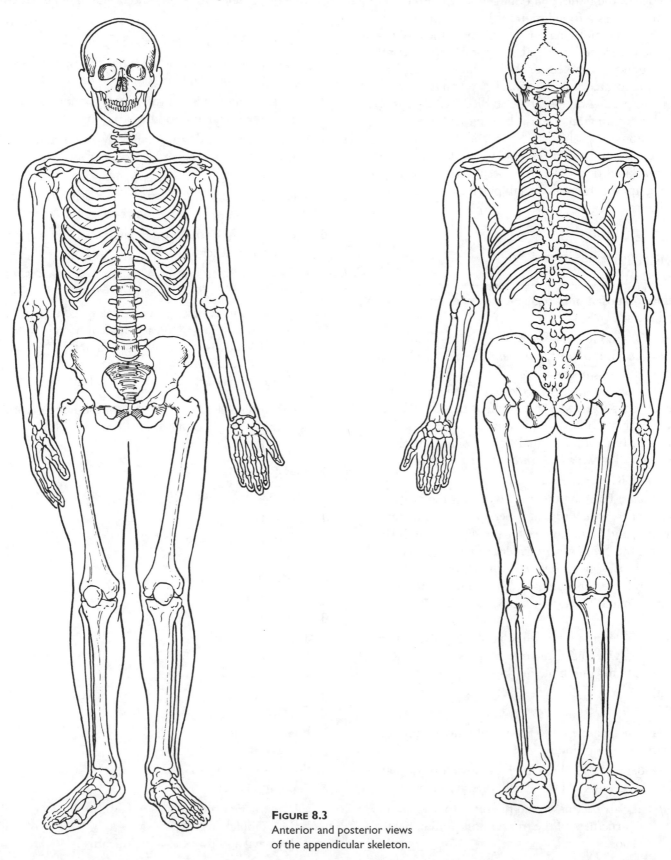

Figure 8.3
Anterior and posterior views
of the appendicular skeleton.

EXERCISES

Knowledge of the bones and bone markings of the skeletal system serves as a foundation for later chapters. For example, the radial and ulnar arteries parallel the radius and the ulna, and the frontal, parietal, temporal, and occipital lobes of the brain are named for the cranial bones under which they are located. In addition, skeletal anatomy is useful for interpreting radiographic studies (i.e., X-rays, CT scans), and in locating anatomical landmarks for procedures such as injections.

The exercises in this unit ask you to identify the bones and bone markings of the skeletal system through a cooperative learning approach involving lab partners rather than individual work.

▌ EXERCISE 1: The Skull

The skull is composed of two general types of bones: cranial bones and facial bones.

1. *Cranial bones*: Encasing the brain are eight cranial bones—the *frontal bone*, *parietal bones*, *occipital bone*, *temporal bones*, *sphenoid bone*, and *ethmoid bone*. Only the parietal bones and temporal bones are paired. Together, they form what is known as the *calvaria* (also known as the "skullcap") and the base of the cranial cavity. The base of the cranial cavity contains three indentations known as the *anterior*, *middle*, and *posterior cranial fossae*.

2. *Facial bones*: Forming the framework for the face are 14 cranial bones, which provide openings for ventilation and eating and form cavities for the sense organs. These bones are the *mandible*, *maxillary bones*, *zygomatic bones*, *lacrimal bones*, *palatine bones*, *nasal bones*, *vomer*, and *inferior nasal conchae*. All facial bones except the mandible and the vomer are paired.

To go about learning the bones, you will engage in *cooperative learning*, working with your lab partners to teach each other the bones and bone markings.

Procedure

1. The class is assembled into groups of a minimum of four students; five is optimum.

2. Each student takes a skull and is assigned one of the following five groups of bones and bone markings:

   Group A: Calvaria, base (and the three fossae), and structures of the frontal bone and parietal bones

   Group B: Temporal bone and occipital bone structures

   Group C: Ethmoid bone and sphenoid bone structures

   Group D: Mandible and maxillary bone structures

   Group E: Remainder of the facial bones, orbit, sutural bones, and anterior and posterior fontanels.

3. Each student spends approximately 3 minutes learning the assigned structures on his or her own.

4. Each student spends approximately 1 to 2 minutes teaching the group his or her assigned structures.

5. The assigned structures are rotated clockwise so each student has a new set of structures to learn (the student that was assigned group A will take group B, and so on).

6. Each student spends approximately 2–3 minutes learning the assigned structures.

7. Each student spends approximately 1–2 minutes teaching the group his or her assigned structures.

8. The process is repeated (it begins to speed up significantly at this point) until each student has taught each group of structures once. By the end of this "game," the students will have learned each group of structures and each group will have been presented four times.

The following are listings of bones and bone markings of the skull that you will cover in the above exercise.

Cranial Bones

See Figures 5-8–5.13 of lab atlas.

1. Calvaria

2. Base of cranial cavity
   a. Anterior cranial fossa
   b. Middle cranial fossa
   c. Posterior cranial fossa

3. Frontal bone
   a. Frontal sinuses
   b. Glabella
   c. Supraorbital foramen

4. Parietal bones
   a. Coronal suture
   b. Sagittal suture
   c. Squamous suture
   d. Lambdoid suture

5. Temporal bones
   a. Zygomatic process
   b. Mandibular fossa
   c. External auditory (acoustic) meatus

d. Styloid process

e. Mastoid process with mastoid air cells

f. Carotid canal

g. Jugular foramen

6. Occipital bone

    a. Occipital condyles

    b. Foramen magnum

    c. External occipital protuberance (or superior nuchal line)

7. Sphenoid bone

    a. Body

    b. Greater and lesser wings

    c. Optic foramen (canal)

    d. Superior orbital fissure

    e. Sella turcica

    f. Sphenoid sinus

8. Ethmoid bone

    a. Perpendicular plate

    b. Superior nasal conchae

    c. Middle nasal conchae

    d. Crista galli

    e. Cribriform plate

    f. Ethmoid sinuses

Facial Bones

See Figures 5.8–5.25 of your lab atlas.

1. Mandible

    a. Mandibular condyle

    b. Coronoid process

    c. Mandibular notch

    d. Angle of mandible

    e. Mental foramen

2. Maxilla

    a. Palatine processes

    b. Infraorbital foramen

    c. Zygomatic processes

    d. Maxillary sinuses

3. Palatine bones

4. Zygomatic bones and zygomatic arch

5. Lacrimal bones and lacrimal fossa

6. Nasal bones

7. Vomer

8. Inferior nasal conchae

Other Structures

See Figures 5.4–5.7 and Figure 5.15 of your lab manual.

1. Sutural (Wormian) bones

2. Auditory ossicles

3. Anterior fontanel

4. Posterior fontanel

5. Orbit: formed by the frontal bone, maxillary bone, zygomatic bone, lacrimal bone, palatine bone, ethmoid bone, and sphenoid bone

> ➤ NOTE: YOUR INSTRUCTOR MAY WISH TO OMIT CERTAIN STRUCTURES INCLUDED ABOVE, OR ADD STRUCTURES NOT INCLUDED IN THESE LISTS. LIST ANY ADDITIONAL STRUCTURES BELOW:

✔ Check Your Understanding

1. Orbital fractures are some of the more difficult fractures to fixate. Beyond the delicate nature of the structures within the orbit (e.g., the eyeball), why do you think that orbital fractures would be so difficult to fixate?

2. The bones of the fetal skull are not yet fused, giving the skull many "soft spots," or *fontanels*. Why do you think the bones of the fetal skull are not fused at birth?

2 EXERCISE 2:
Remainder of the Axial Skeleton

The remainder of the axial skeleton consists of the bones of the *vertebral column*, the *ribs*, the *sternum*, and the *hyoid bone*. These are outlined as follows:

1. *Vertebrae*: The 22 total vertebrae are divided into 7 *cervical vertebrae*, 12 *thoracic vertebrae*, and 5 *lumbar vertebrae*. All vertebrae share certain general features, which you labeled in the Pre-Lab Exercises. In addition, vertebrae of different regions share unique features that help you identify a vertebra as cervical, thoracic, or lumbar. See Figures 5.28–5.29 in your lab atlas for reference.

    a. *Cervical vertebrae* share the following common features:

    (1) *Transverse foramina*: Found only in cervical vertebrae, these holes in the transverse processes permit the passage of blood vessels called the *vertebral artery* and *vein*.

    (2) *Bifid spinous processes*: This means that the spinous processes of these vertebrae are often forked.

    Two cervical vertebrae are named separately from the others because of their unique features.

    (1) *Atlas* (C1): The first cervical vertebra, it articulates with the occipital bone at the occipital condyles. It is easily identified because it has a large vertebral foramen, no body, and no spinous process.

    (2) *Axis* (C2): The second cervical vertebra, it articulates with the atlas to form the *atlantoaxial joint*. It is also easily identified by a superior projection called the *dens* (or the *odontoid process*). The dens fits up inside the atlas and allows rotation of the head.

    b. *Thoracic vertebrae* share the following common features:

    (1) The spinous processes are thin and point inferiorly.

    (2) All 12 thoracic vertebrae have two *articular facets*, which articulate with the ribs (there are 12 pairs of ribs).

    (3) All have triangular vertebral foramina.

    (4) If you look at a thoracic vertebra from the posterior side, it looks like a giraffe (seriously!).

    c. *Lumbar vertebrae* share the following common features:

    (1) All have a large, block-like body.

    (2) The spinous processes are thick and point posteriorly.

    (3) If you look at a lumbar vertebra from the posterior side, it looks like a moose (really!).

2. *Ribs*: There are three types of ribs, classified according to their attachment to the sternum. See Figures 5.35–5.37 for reference.

    a. *True ribs*: Ribs 1–7 are considered true ribs because they attach directly to the sternum by their own cartilage.

    b. *False ribs*: Ribs 8–10 are false ribs because they attach to the cartilage of the true ribs rather than directly to the sternum.

    c. *Floating ribs*: Ribs 11–12 are called floating ribs because they lack an attachment to the sternum.

3. *Sternum*: The sternum is divided into three parts:

    a. the upper *manubrium,*

    b. the middle *body,* and

    c. the lower *xiphoid process.*

4. *Hyoid bone*: The hyoid bone is often classified as a skull bone, although it does not articulate with any skull bone or any other bone. It is held in place in the superior neck by muscles and ligaments, and it helps to form part of the framework for the larynx (voice box). It also serves as an attachment site for the muscles of the tongue and aids in swallowing. When a person is manually choked, the hyoid bone is often broken; however, I would not recommend testing this on your lab partner!

The following is a list of bones and bone markings of the axial skeleton that will be covered in this lab. See Figures 5.26–5.37 in your lab atlas.

> ➤ NOTE: WE WILL COVER THE REMAINING STRUCTURES OF THE AXIAL SKELETON WITH A COOPERATIVE LEARNING EXERCISE IN THE NEXT SECTION.

1. Vertebrae
   a. Body
   b. Spinous process
   c. Vertebral foramen
   d. Vertebral arch
   e. Transverse processes

2. Cervical vertebrae
   a. Atlas
   b. Axis with dens
   c. Transverse foramina

3. Thoracic vertebrae
   a. Articular facets

4. Lumbar vertebrae

5. Sacrum
   a. Median sacral crest
   b. Posterior sacral foramina

6. Coccyx

7. Ribs
   a. True ribs
   b. False ribs
   c. Floating ribs
   d. Tubercle
   e. Head
   f. Neck
   g. Shaft

8. Sternum
   a. Manubrium
   b. Body
   c. Xiphoid process

9. Hyoid bone

3 EXERCISE 3:
The Appendicular Skeleton _____

The structures of the appendicular skeleton are divided into the following classes:

1. *Pectoral girdle*: bones that frame the shoulder, and include the *scapula* and the *clavicle*.

2. *Upper limb*: consists of the *arm*, which contains only one bone, the *humerus*; the *forearm*, which contains the *radius* and the *ulna*; and the wrist and hand, composed of 8 *carpals*, 5 *metacarpals*, and 14 *phalanges*.

3. *Pelvic girdle*: contains the bones of the hip: the *ilium*, the *ischium*, and the *pubis*. In this exercise, we will examine the differences between the male and female pelvises. In general, the male and female pelvis can be distinguished by the features listed in the chart below:

FEATURE	FEMALE PELVIS	MALE PELVIS
Pelvic inlet shape	Wider and oval-shaped	Narrower and heart-shaped
Pubic arch	Wide angle	Narrow angle
Acetabulae	Farther apart	Closer together
Ischial tuberosities	Everted	Inverted
Coccyx	Straighter, more movable	Curved anteriorly, less movable

4. *Lower limb*: this includes the *thigh*, which contains only the *femur*; the *patella*; the *leg*, which is composed of the *tibia* and *fibula*; and the *foot* and *ankle*, composed of 7 *tarsals*, 5 *metatarsals*, and 14 *phalanges*.

We will follow essentially the same procedure here as we did for Exercise 1, but with different groups of bones. Again, this is a cooperative learning exercise.

Procedure

1. The class is assembled into groups with a minimum of four students; five is optimum.

2. Each student is assigned one of the following five groups of bones and bone markings:

   Group A: Cervical, thoracic, and lumbar vertebrae and vertebral markings; ribs and rib markings; sternum structures and the hyoid bone

   Group B: Scapula structures, clavicle structures, humerus structures

   Group C: Radius and ulna markings, carpals, metacarpals, and phalanges

   Group D: Ilium, ischium, and pubis structures, and the difference between the male and female pelvises

   Group E: Femur, tibia, fibula structures, tarsals, metatarsals, phalanges.

3. Each student spends approximately 3 minutes learning his or her assigned structures independently.

4. Each student then teaches the group his or her assigned structures.

5. The assigned structures are rotated clockwise so each student has a new set of structures to learn (the student who was assigned to group A will move to group B, and so on).

6. The students spend approximately 2–3 minutes learning their new assigned structures.

7. Each student then teaches the group his or her group of structures.

8. The process is repeated (it begins to speed up significantly at this point) until each student has been taught each group of structures once. So, by the end of this exercise, the students all will have learned each group of structures, and each group will have been presented five times.

The following is a list of bone and bone markings of the appendicular skeleton that you will cover in the above exercise:

Pectoral Girdle

See Figures 6.3 and 6.5 in your lab atlas.

1. Clavicle
   a. Acromial end
   b. Sternal end
   c. Conoid tubercle
2. Scapula
   a. Acromion
   b. Spine
   c. Coracoid process
   d. Glenoid cavity (fossa)
   e. Lateral, medial, and superior borders

Upper Limb

See Figures 6.6–6.14 in your lab atlas.

1. Humerus
   a. Head
   b. Greater tubercle
   c. Lesser tubercle
   d. Intertubercular groove
   e. Deltoid tuberosity
   f. Medial and lateral epicondyles
   g. Capitulum
   h. Trochlea
   i. Olecranon fossa
   j. Coronoid fossa
2. Radius
   a. Head
   b. Neck
   c. Radial tuberosity
   d. Styloid process
3. Ulna
   a. Olecranon process
   b. Coronoid process
   c. Trochlear notch
   d. Styloid process
   e. Radial notch
4. Carpals
   a. Scaphoid
   b. Lunate
   c. Triquetrum
   d. Pisiform
   e. Trapezium
   f. Trapezoid

g. Capitate

h. Hamate

5. Metacarpals

6. Phalanges

Pelvic Girdle

See Figures 6.15–6.18 in your lab atlas.

1. Ilium
   a. Iliac crest
   b. Anterior superior iliac spine
   c. Anterior inferior iliac spine
   d. Posterior superior iliac spine
   e. Posterior inferior iliac spine
   f. Greater sciatic notch

2. Ischium
   a. Ischial tuberosity
   b. Lesser sciatic notch
   c. Ischial spine
   d. Obturator foramen

3. Pubis
   a. Superior and inferior rami
   b. Pubic arch
   c. Pubic symphysis

4. Acetabulum

5. Male and female pelvises

Lower Limb

See Figures 6.19–6.28 in your lab atlas.

1. Femur
   a. Head
   b. Neck
   c. Greater trochanter
   d. Lesser trochanter
   e. Linea aspera
   f. Lateral and medial epicondyles
   g. Medial and lateral condyles
   h. Intercondylar groove (fossa)

2. Patella

3. Tibia
   a. Intercondylar eminence
   b. Medial and lateral condyles
   c. Tibial tuberosity
   d. Medial malleolus

4. Fibula
   a. Head
   b. Lateral malleolus

5. Tarsals
   a. Talus
   b. Calcaneus
   c. Navicular
   d. Cuboid
   e. Cuneiforms
   f. Metatarsals
   g. Phalanges

> ➤ **NOTE:** YOUR INSTRUCTOR MAY WISH TO OMIT CERTAIN STRUCTURES INCLUDED ABOVE, OR ADD STRUCTURES NOT INCLUDED IN THESE LISTS. LIST ANY ADDITIONAL STRUCTURES BELOW:

✔ Check Your Understanding

Explain the reasons for the anatomical differences in the male and female pelvises.

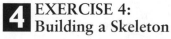

4 EXERCISE 4: Building a Skeleton _____

Obtain a set of disarticulated bones (real bones are best), and assemble the bones into a full skeleton. (If you have an articulated vertebral column, go ahead and use it, as assembling a vertebral column is quite time-consuming.) Be certain to keep the skeleton in anatomical position.

9. Articulations

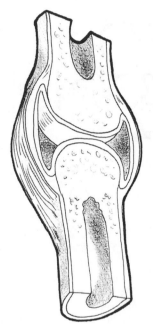

■ **OBJECTIVES**

Once you have completed this unit, you should be able to:

1. Classify joints based upon structure and function.
2. Identify examples of the different types of joints.
3. Identify structures associated with synovial joints.
4. Classify synovial joints according to range of motion.
5. Identify structures of the knee joint.
6. Observe and dissect fresh joint specimens.
7. Demonstrate and describe motions allowed at synovial joints.

■ **MATERIALS**

● Anatomical models: representative models of fibrous, cartilaginous, and synovial joints, the knee, hip, elbow, and shoulder joints, and articulated skeletons
● Fresh pig or chicken joints
● Dissection trays and kits

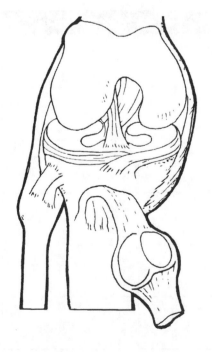

PRE-LAB EXERCISES

Prior to coming to lab, complete the following exercises, using your lab atlas and textbook for reference.

1 PRE-LAB EXERCISE 1: Typical Synovial Joints

Use colored pencils to color-code and label the diagram of a typical synovial joint in Figure 9.1, using the terms from Exercise 2. Refer to Chapter 7 in your lab atlas for assistance.

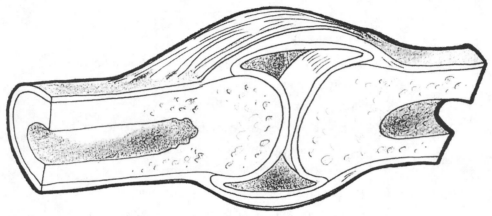

FIGURE 9.1
Diagram of a typical synovial joint.

2 PRE-LAB EXERCISE 2: Subclasses of Synovial Joints

Synovial joints are those with true joint cavities. They allow the most motion of any type of joint, but the amount of motion varies depending upon the structure of the joint. Based upon these structural differences, synovial joints are divided into subclasses, each with differing ranges of motion. Fill in the following chart with definitions and examples of each subclass of synovial joint, using Chapter 7 in your lab atlas as a guide.

SUBCLASS	DEFINITION	EXAMPLE
Plane		
Hinge		
Pivot		
Condyloid		
Saddle		
Ball and socket		

3 PRE-LAB EXERCISE 3:
The Knee Joint_____

Use colored pencils to color-code and label the diagram of the knee joint given in Figure 9.2, using the terms from Exercise 2. See Figures 7.22–7.26 and 7.29 in your lab atlas for reference.

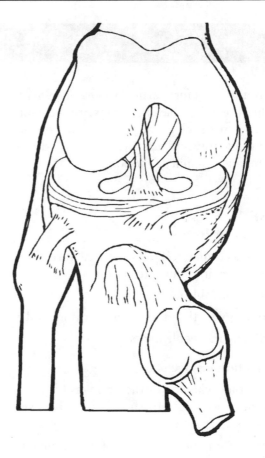

FIGURE 9.2
Diagram of knee joint.

4 PRE-LAB EXERCISE 4:
Motions of Synovial Joints _____

Fill in the following chart with descriptions of the common motions allowed at synovial joints and examples of joints that allow this motion, using Table 7.1 in your lab atlas for reference.

MOTION	DEFINITION	EXAMPLE
Flexion		
Extension		
Adduction		
Abduction		
Circumduction		
Rotation		
Inversion		
Eversion		
Pronation		
Supination		
Elevation		
Depression		
Opposition		

EXERCISES

Most bones in the body form articulations, or joints, with another bone (notable exceptions are the hyoid bone and sesamoid bones such as the patella). The following exercises ask you to identify and classify joints according to structure, function, and amount of motion.

1 EXERCISE 1: Classification of Joints

Joints are classified according to both their structure and their function. Functionally, joints are classed as:

1. *Synarthroses*: immovable joints,
2. *Amphiarthroses*: joints that allow some motion, and
3. *Diarthroses*: freely moveable joints.

   Structural classifications include:

1. *Fibrous joints*: joints classified structurally as fibrous joints consist of bones joined by short connective tissue fibers; most are synarthroses.
2. *Cartilaginous joints*: joints consisting of bones united by cartilage rather than fibrous connective tissue; most are amphiarthrotic.
3. *Synovial joints*: joints that have a true joint cavity and consist of two bones whose articular ends are covered with hyaline cartilage. The joint cavity is lined by a synovial membrane, which secretes a watery fluid similar in composition to blood plasma without the proteins. The fluid bathes the joint to permit frictionless motion. We will discuss synovial joints further in Exercise 2.

Fill in the following chart with the structural classification of the listed joints. Then examine and manipulate the joint to determine the amount of motion allowed at the joint, and therefore the functional classification. See Chapter 7 in your lab atlas for assistance.

2 EXERCISE 2: Synovial Joints

Synovial joints are those that have a fluid-filled cavity lined by a synovial membrane. Features common to synovial joints include:

- *Joint capsule*: The joint capsule is made of dense irregular connective tissue proper and serves to provide strength and structural reinforcement for the joint. It is lined by the *synovial membrane*, which secretes synovial fluid into the joint cavity. This fluid lubricates the joint.
- *Articular cartilage*: The articulating ends of the bones are covered with articular cartilage, which usually is hyaline cartilage. The cartilage provides a smooth, frictionless surface on which the bones slide.
- *Ligaments*: To further strengthen the joint, the bones are held together by a series of ligaments. Some ligaments are within the joint cavity (called extrinsic ligaments); others are embedded in the capsule (intrinsic ligaments).
- *Articular discs*: Also known as *menisci*, these fibro-cartilage pads improve the fit of two bones to prevent dislocation.

Synovial joints typically are surrounded by tendons that move the bones involved in the joint. To prevent the tendons from rubbing on the joint, two features are often present:

1. *Tendon sheath*, a sheath of connective tissue that surrounds the tendon, in which it can slide with a minimum of friction.
2. *Bursa*, a fluid-filled sac between the tendon and the joint, which also lessens friction.

Structures of Synovial Joints

Using a fresh joint specimen (chicken or beef joints or pigs' feet work well), identify the following structures. If fresh specimens are not available, use anatomical models instead.

JOINT	STRUCTURAL CLASSIFICATION	AMOUNT OF MOTION	FUNCTIONAL CLASSIFICATION
Intervertebral joint			
Shoulder (gleno-humeral) joint			
Intercarpal joint			
Coronal suture			
Pubic symphysis			
Interphalangeal joint			

1. Joint cavity
2. Joint capsule
3. Articular cartilage
4. Articular discs (menisci)
5. Synovial membrane
6. Synovial fluid
7. Bursae
8. Tendon with tendon sheath
9. Ligaments
   a. Intrinsic
   b. Extrinsic

Types of Synovial Joints

Recall from the Pre-Lab Exercises that the amount of motion of synovial joints depends upon the type or subclass of joint. This motion is described conventionally in terms of an invisible *axis* about which the bone moves.

● *Nonaxial joints*: joints in which neither bone moves around the other. An example is found at the vertebrocostal joint.

● *Uniaxial joints*: joints that allow motion in one plane or direction only. A classic example is the elbow, which permits only flexion and extension.

● *Biaxial joints*: joints that allow motion in two planes. An example is the wrist, which allows both flexion / extension and abduction / adduction.

● *Multiaxial joints*: joints that allow motion in multiple planes, and therefore have the greatest range of motion. A good example is the shoulder joint.

Now let's put these terms together with the terms from Pre-Lab Exercise 2. For each of the following joints, first list in the chart below the subclass of synovial joint. Then obtain an articulated skeleton so you can manipulate each joint and determine if the joint is nonaxial, uniaxial, biaxial, or multiaxial.

JOINT	TYPE	RANGE OF MOTION
Intercarpal joint		
Proximal radio-ulnar joint		
Radiocarpal joint		
Thumb carpo-metacarpal joint		
Interphalangeal joint		
Knee joint		
Atlantoaxial joint		
Hip joint		

✔ Check Your Understanding

1. Compare the structure of the hip joint and the shoulder joint. Which do you think would be more likely to dislocate? Why?

2. The term "double-jointed" describes individuals who have an abnormally large range of motion in a given joint. The cause for this excess range of motion isn't the presence of a second joint, as the name implies, but instead laxity of the ligaments surrounding the joint. Why would laxity of ligaments lead to an increased range of motion at a joint? What may be some consequences of ligamentous laxity?

3. A common result of trauma to a joint is post-traumatic arthritis, in which cartilage and other structures in the joint are damaged. Considering the blood supply of joints, explain why they tend to heal poorly when injured, leading to permanent changes.

Knee Joint

The knee joint is a modified hinge joint stabilized by intracapsular and extracapsular ligaments that reinforce the joint, and by the medial and lateral menisci, which improve the fit of the femur on the tibia. Four important extracapsular ligaments are described below:

● *Anterior cruciate ligament (ACL)*: extends from the anterior tibial plateau to the lateral femoral condyle; its function is to prevent hyperextension of the knee.

● *Posterior cruciate ligament (PCL)*: extends from the posterior tibial plateau to the medial femoral condyle; its function is to prevent posterior displacement of the tibia on the femur.

● *Medial collateral* and *lateral collateral ligaments*: extend from the medial tibia and the lateral fibula to the femur, respectively. Their function is to resist varus and valgus (medial and lateral) stresses.

Identify the following structures of the knee joint on anatomical models or fresh specimens. For reference, see Figures 7.22–7.26 and 7.29 in your lab atlas, as well as Pre-Lab Exercise 2.

1. Fibrous capsule

2. Ligaments
   a. Lateral collateral ligament
   b. Medial collateral ligament
   c. Anterior cruciate ligament (ACL)
   d. Posterior cruciate ligament (PCL)
   e. Oblique popliteal ligament
   f. Patellar ligament (tendon)

3. Menisci
   a. Medial meniscus
   b. Lateral meniscus

4. Medial and lateral femoral condyles

5. Tibial plateau

Testing the Integrity of the ACL and PCL

One of the more common injuries encountered in sports is damage to the ACL and/or the PCL. To assess the integrity of the ACL and PCL on the sidelines, tests known as the *anterior drawer* and *posterior drawer* are performed, respectively. A normal result of both tests is a minimum of motion when the tibia is moved anteriorly and posteriorly.

1. Have your partner sit with his or her knees bent at 90 degrees and relaxed.

2. Grasp your partner's leg with both hands around the proximal tibia and fibula.

3. Gently pull the leg anteriorly, being careful not to extend the knee joint in the process.

Amount of motion:

Which ligament was being assessed (ACL or PCL)?

What would you have expected to see if the test were abnormal?

4. Now gently push the leg posteriorly, being careful not the flex the knee joint.

Amount of motion:

Which ligament was being assessed (ACL or PCL)?

What would you have expected to see if the test were abnormal?

3 EXERCISE 3: Motions of Synovial and Cartilaginous Joints__

Each time you move your body in a seemingly routine fashion (such as walking or climbing stairs), you are creating motion at a tremendous number of joints. This exercise allows you to determine which joints you are moving with some commonly performed actions, and which motions are occurring at each joint.

Before you begin this exercise, review each motion in Table 7.1 of your lab atlas and Pre-Lab Exercise 4, and examine the figures on pages 62–64 in your atlas. After completing this review, have your partner perform each of the following actions. Watch carefully as the actions are being performed, and list the joints that are in motion.

Ask your instructor if he/she wants you to use the technical or the common name for each joint (e.g., glenohumeral versus shoulder joint). Some of the joints will be obvious, such as the hip and knee joints. Others are less obvious and easily overlooked, such as the radioulnar joint, the fingers and toes, and the sacroiliac joint.

Once you have listed the joints that are being used, determine which motions are occurring at each joint (refer to the chart in Pre-Lab Exercise 4 for a list of terms). Keep in mind the type and the range of motion of each joint as you answer each question.

1. Climbing the stairs:

Joints moving: Motions occurring:

2. Doing jumping jacks:

Joints moving: Motions occurring:

3. Answering the telephone:

Joints moving: Motions occurring:

4. Jumping rope:

Joints moving: Motions occurring:

10. Muscle Tissue

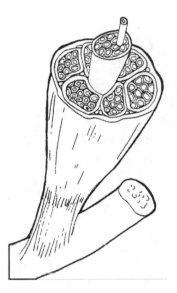

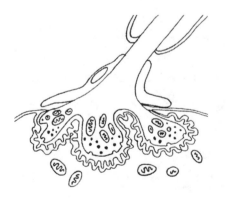

■ **OBJECTIVES** _____

Once you have completed this unit, you should be able to:

1. Describe the microscopic anatomy of skeletal muscle cells.

2. Identify structures of skeletal muscle cells on anatomical models and microscope slides.

3. Identify structures associated with the neuromuscular junction on anatomical models and microscope slides.

4. Distinguish skeletal muscle tissue from smooth and cardiac muscle tissue.

■ **MATERIALS** _____

● Anatomical models: skeletal muscle cell, neuromuscular junction

● Microscope slides: skeletal, smooth, and cardiac muscle tissue, and neuromuscular junction

● Colored pencils

● Oil immersion microscope for demonstration purposes

PRE-LAB EXERCISES

Prior to coming to lab, complete the following exercises, using Chapter 8 in your atlas for reference.

1 PRE-LAB EXERCISE 1: Skeletal Muscle Anatomy

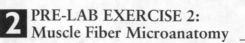

Label Figure 10.1 with the terms relating to skeletal muscle given in Exercise 1. Use your colored pencils to color-code the structures as you label them.

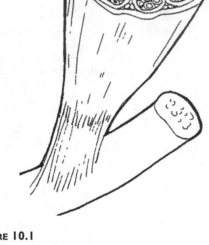

FIGURE 10.1
Diagram of skeletal muscle anatomy.

2 PRE-LAB EXERCISE 2: Muscle Fiber Microanatomy

Label Figure 10.2 with the terms relating to muscle fiber microanatomy from Exercise 1. Note that all structures may not be visible on the diagrams. Use your colored pencils to color-code the structures as you label them. See Figure 8.2 in your lab atlas for reference.

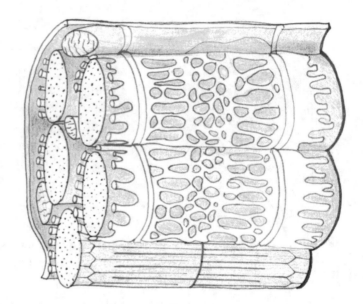

FIGURE 10.2
Diagram of myofibrils.

3 PRE-LAB EXERCISE 3: Neuromuscular Junction

Label Figure 10.3 with the terms relating to the skeletal muscle cell neuromuscular junction from Exercise 2. Note that all structures may not be visible on the diagram. Use your colored pencils to color-code the structures as you label them.

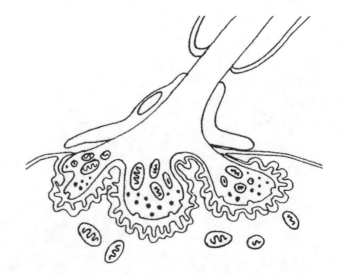

FIGURE 10.3
Diagram of skeletal muscle cell neuromuscular junction.

4 PRE-LAB EXERCISE 4:
Types of Muscle Tissue

Fill in the following chart with the differences between skeletal, smooth, and cardiac muscle tissue (see Unit 5).

MUSCLE TYPE	STRIATED OR NONSTRIATED	ONE OR MULTIPLE NUCLEI	SIZE AND SHAPE OF CELLS	VOLUNTARY OR INVOLUNTARY	SPECIAL FEATURES
Cardiac muscle					
Skeletal muscle					
Smooth muscle					

EXERCISES

1 EXERCISE 1:
Skeletal Muscle

An understanding of muscle tissue is fundamental to understanding muscle physiology. Although much of the emphasis in this unit is on the structures of skeletal muscle tissue, keep in mind that there are three structurally distinct types of muscle tissue: skeletal muscle, cardiac muscle, and smooth muscle.

Skeletal muscle cells, also called *muscle fibers*, are long, cylindrical cells surrounded by a connective tissue sheath called the *endomysium*, which overlies their *sarcolemma* (cell membrane). They are arranged in bundles called *fascicles*, which contain thousands of muscle fibers and are surrounded by another connective tissue sheath, the *perimysium*. The muscle as a whole, composed of groups of fascicles, is covered by the *epimysium*, which blends with the thick, superficial fascia that binds muscles into groups. The epimysium also blends with the fibers of tendons and aponeuroses, which connect the muscle to bones or soft tissue.

Microscopically, muscle cells are composed of smaller units called *myofibrils*, which themselves are made of smaller protein subunits called *myofilaments*. The arrangement of myofilaments within the myofibrils is what gives skeletal muscle its characteristic striated appearance. The two types of myofilaments are: (1) *thick filaments*, composed of the protein *myosin*, and (2) *thin filaments*, composed mostly of the protein *actin*.

The dark regions of the striations, called *A bands*, are dark because of the overlap of the thick and thin filaments (see Figure 8.2 in your lab atlas). The light regions, called *I bands*, appear light because they contain only thin filaments. Bisecting the A band is a lighter region called the *H zone* (which itself is bisected by a line called the *M line*), and bisecting the I band is a dark line called the *Z disc*. (At this point, you may wish to take a deep breath and read this again.)

The terms of this alphabet soup are used to describe the fundamental unit of contraction, the *sarcomere*, defined as the space from one Z-disc to the next Z-disc. Each myofibril consists of repeating sarcomeres. During a muscle contraction, the thick and thin filaments slide past one another, which narrows the I band and the H zone (the width of the A band remains unchanged).

Additional microscopic features of muscle fibers include inward extensions of the sarcolemma called transverse tubules (usually shortened to simply *T-tubules*). T-tubules are located at the Z-discs and are flanked on either side by two terminal cisternae, part of the calcium-storing *sarcoplasmic reticulum* (a modified smooth endoplasmic reticulum). A T-tubule combined with two terminal cisternae is called a *triad*.

Identify the following structures of skeletal muscle tissue on anatomical models and charts, using Figure 8.2 in your lab atlas and the Pre-Lab Exercises for reference.

Skeletal Muscle Anatomy

1. Connective tissue coverings
   a. Epimysium
   b. Perimysium
   c. Endomysium
2. Fascicle
3. Muscle fiber (cell)
4. Tendon
5. Aponeurosis

Muscle Fiber Microanatomy

1. Sarcolemma
2. Sarcomere
   a. A band
   b. I band
   c. H zone
   d. Z disc
   e. M line
3. Transverse tubules
4. Sarcoplasmic reticulum
5. Terminal cisternae
6. Triad

Model Inventory

As you examine the anatomical models and diagrams in lab, list them on the inventory below and state which structures you are able to locate on each model.

MODEL/DIAGRAM	STRUCTURES IDENTIFIED

EXERCISE 2:
The Neuromuscular Junction _____

When a skeletal muscle contracts, a specific number of fibers is activated to produce a contraction of a specific strength, a phenomenon called *recruitment*. This is possible because each skeletal muscle cell has its own nerve supply. The place where the nerve meets the muscle cell is called the *neuromuscular junction*.

The neuromuscular junction consists of three parts:

1. *Axonal terminal*: Each motor nerve splits into many axonal terminals, each of which contacts one muscle fiber. The axonal terminal has *synaptic vesicles* in its cytosol, which contain the neurotransmitter *acetylcholine*.

2. *Synaptic cleft*: The axonal terminal doesn't come into direct contact with the muscle fiber; instead there is a space between the two called the synaptic cleft. When a nerve impulse reaches the axonal terminal, exocytosis of the synaptic vesicles is triggered and the acetylcholine gets released to diffuse across the synaptic cleft.

3. *Motor end plate*: This is a specialized region of the sarcolemma that contains acetylcholine receptors. When acetylcholine binds to these receptors, an action potential is started. The action potential propagates through the sarcolemma, diving into the fiber along the T-tubules. As it spreads past the terminal cisternae, calcium is released, which begins the sequence of events for a muscle contraction.

Parts of the Neuromuscular Junction

Identify the following parts of the neuromuscular junction on models and diagrams, using the Pre-Lab Exercises for reference. As you identify the structures, list them on the model inventory from Exercise 1.

1. Motor nerve
2. Axonal terminal
3. Synaptic vesicles
4. Synaptic cleft
5. Motor end plate
6. Acetylcholine receptors

Histology of the Neuromuscular Junction

Obtain a prepared slide of a neuromuscular junction. Use your colored pencils to draw what you see, and label your drawing with the following terms:

1. Motor nerve
2. Axonal terminal
3. Skeletal muscle fiber
4. Striations

✔ Check Your Understanding

1. A current trend in cosmetic surgery involves use of the drug Botox®, in which the toxin from the bacterium *Clostridium botulinum* is injected subdermally to minimize fine lines and creases in the face. This toxin works by preventing motor neurons from releasing acetylcholine. How would this produce the desired cosmetic results? What would happen if this toxin were absorbed systemically?

2. *Myasthenia gravis* is an autoimmune disease in which the immune system produces antibodies that attack the acetylcholine receptor. What symptoms would you expect with this disease? Explain.

3 EXERCISE 3: Muscle Tissue _____

Recall from the Pre-Lab Exercises the three primary types of muscle tissue:

1. *Skeletal muscle*: typically found along the skeleton, although also found in other locations, such as the superior part of the esophagus and around the pharynx. This is the only type of muscle that is under voluntary control, and requires stimulation by a motor nerve to contract.

2. *Cardiac muscle*: found only in the myocardium of the heart. Its structure is similar to that of skeletal muscle (with some notable differences that we will discuss momentarily), but it is involuntary and autorhythmic—cardiac muscle controls its own pace of contraction and relaxation via specialized cells called *pacemaker cells*.

3. *Smooth muscle*: found lining all hollow organs. All smooth muscle is involuntary and most requires extrinsic innervation to contract, although some parts of the gastrointestinal tract contain pacemaker cells.

Distinguishing Muscle Types

Look for the following distinguishing features of the three types of muscle on microscope slides:

● *Striations*: If striations (lines) are present, it is either skeletal or cardiac muscle. If there are no striations, it is smooth muscle. Recall that the alternating dark and light bands in the myofibrils are what create the striations.

● *Shape of the cells*: Skeletal muscle cells are long and thin, with the nuclei shoved up next to the sarcolemma. Cardiac muscle cells are short and fat, with their nuclei interspersed with the myofibrils. Smooth muscle cells are long, thin, and spindle-shaped. The single nucleus is located in the middle of the cell.

● *Special features*:
  • Skeletal muscle cells are multinucleated. As you look at skeletal muscle tissue, keep in mind that each long cylinder is one cell. So when you see the nuclei up against the sarcolemma, those are multiple nuclei within one cell. Both smooth and cardiac muscle are uninucleated.

  • Cardiac muscle cells have a special feature called *intercalated discs*, located between individual cells. The discs appear as black lines that are oriented parallel to the striations. They contain desmosomes and gap junctions to allow communication between cardiac cells so the heart beats as a unit.

  • Smooth muscle tissue typically is found on a slide with other types of tissue. Because it lines hollow organs, you generally can see the other tissues in the wall of the hollow organ (usually epithelial and connective tissue). You also may see blood vessels scattered throughout the tissue (see Figure 3.38 in your lab atlas). If you're unsure, scroll through the slide to see if you can find any additional tissue types.

Examine prepared slides of skeletal, cardiac, and smooth muscle. The features of each are best seen on high power. For each slide, draw what you see, and label your drawing with the terms below. Use Figures 3.34–3.39 in your atlas for reference.

1. *Skeletal muscle*
   a. Striations
      (1) A band
      (2) I band
   b. Sarcolemma
   c. Nuclei
   d. Endomysium

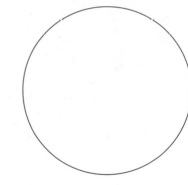

2. *Cardiac muscle*
   a. Striations
      (1) A band
      (2) I band
   b. Sarcolemma
   c. Nucleus
   d. Intercalated disc

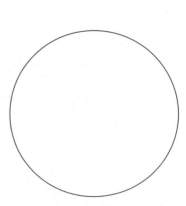

3. *Smooth Muscle*
   a. Nucleus
   b. Blood vessels
      (or other tissues)

Oil Immersion

It is often difficult to distinguish all of the features of skeletal muscle microanatomy (particularly the parts of the sarcomere) on a compound light microscope; however, many of these features are easily seen using an oil-immersion lens. Recall from Unit 3 that an oil immersion lens gives a total of 400X magnification, using oil as a refractive medium. If available, use the oil-immersion lens to view the slide of skeletal muscle. Draw what you see, and label the features listed below.

- Sarcomere
- A band
- I band
- Z disc
- H zone
- M line

✔ Check Your Understanding

1. With respect to cardiac muscle, how does the structure of the cells and intercalated discs (the "form") follow the function of cardiac tissue?

2. Would you expect to see sarcomeres in smooth muscle? Why or why not?

3. Why is skeletal muscle the only type of muscle that is multinucleated?

11. Muscle Anatomy

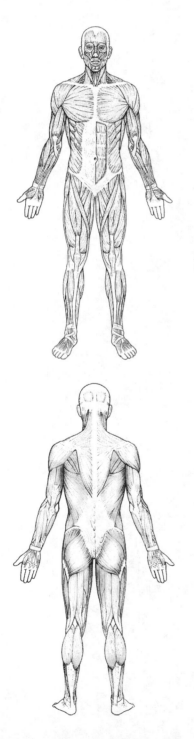

■ **OBJECTIVES**

Once you have completed this unit, you should be able to:

1. Identify muscles of the upper and lower limbs, trunk, head, and neck on anatomical models.

2. Describe the origin, insertion, and action of selected muscles.

3. List the muscles required to perform common movements.

■ **MATERIALS**

● Anatomical models: musculature of the upper and lower limb, trunk, head, and neck

● Colored pencils

● Small skeletons

● Modeling clay in five colors

PRE-LAB EXERCISES

Prior to coming to lab, complete the following exercises, using Chapter 8 in your lab atlas and your textbook for reference.

1 PRE-LAB EXERCISE 1: Muscles of the Human Body

Label and color-code Figure 11.1 with the names of muscles in Exercise 1. Note that not all muscles will be visible in these figures. See Chapter 8 in your lab atlas for reference.

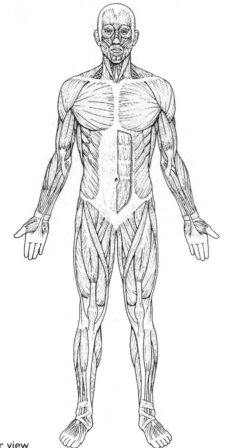

Anterior view

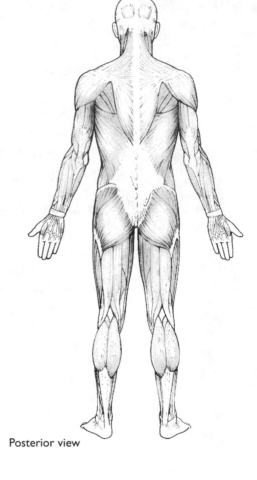

Posterior view

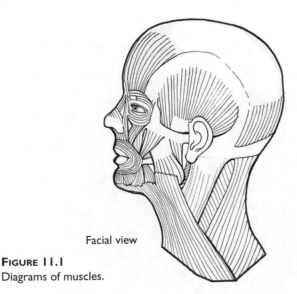

Facial view

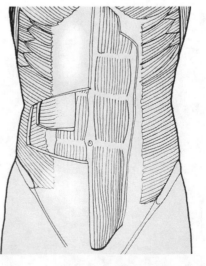

Trunk

FIGURE 11.1
Diagrams of muscles.

2 PRE-LAB EXERCISE 2: Muscle Origins, Insertions, and Actions _____

Fill in the following chart with the origins, insertions, and *main* actions for selected muscles. You may notice in your text and lab atlas that the origin and insertion of a muscle are often extensive. For example, the latissimus dorsi muscle originates from T7–L5, the lower three or four ribs, the inferior angle of the scapula, and the thoracolumbar fascia.

This is obviously quite a bit to write and try to remember, so I suggest shortening it to something that makes sense to you (e.g., "lower back and scapula").

If your instructor wishes you to learn the more technical version, learning a simplified version first may be helpful. You also will notice that most muscles have more than one action, which you may wish to simplify as well. Chapter 8 of your lab atlas contains tables that you may use for reference.

MUSCLE	ORIGIN	INSERTION	ACTION(S)
Sternocleidomastoid m.			
Trapezius m.			
Pectoralis minor m.			
Rectus abdominis m.			
External obliques m.			
Internal obliques m.			
Serratus anterior m.			
Transversus abdominis m.			
Erector spinae m.			
Deltoid m.			
Pectoralis major m.			
Latissimus dorsi m.			
Biceps brachii m.			
Triceps brachii m.			
Brachialis m.			
Brachioradialis m.			
Iliopsoas m.			
Gluteus maximus m.			
Gluteus medius m.			
Sartorius m.			
Tensor fasciae latae m.			
Rectus femoris m.			
Vastus medialis m.			
Vastus intermedius m.			
Vastus lateralis m.			
Adductor group			
Gracilis m.			
Biceps femoris m.			
Semitendinosus m.			
Semimembranosus m.			
Gastrocnemius m.			
Soleus m.			
Tibialis anterior m.			

EXERCISES

The human body has nearly 700 skeletal muscles, ranging dramatically in size and shape from the large and triangular trapezius muscle to the smaller, circular oribcularis oris muscle to the tiny corrugator supercilii muscle. Luckily for you, we will only be learning about 70 muscles rather than the full 700.

A muscle begins at its origin, which is often the more stationary part, and extends to attach to its insertion, which is generally the part that the muscle moves. We are most familiar with muscles that insert into and move bones. Many muscles, however, insert into structures other than bones. Examples include the diaphragm, which inserts into its own central tendon, and the muscles of facial expression, which insert into skin or other muscles.

▌ EXERCISE 1: ▌ Identification of Skeletal Muscles _____

Identify the following muscles on models and diagrams using Pre-Lab Exercise 1, your textbook, and Chapter 8 in your lab atlas as a guide. As you identify the muscles, pay attention to their origins and insertions. Note that the muscles are listed according to the part of the body that is moved rather than the anatomical location (e.g., the latissimus dorsi muscle is included with the muscles of the upper limb rather than the muscles of the trunk).

Muscles of the Head and Neck

See Figures 8.5–8.7 in your lab atlas.

1. Muscles of facial expression
   a. Epicranius m. (frontalis m. and occipitalis m.)
   b. Orbicularis oculi m.
   c. Zygomaticus m.
   d. Buccinator m.
   e. Orbicularis oris m.
2. Muscles of mastication
   a. Temporalis m.
   b. Masseter m.
3. Muscles of the head and neck
   a. Platysma m.
   b. Sternocleidomastoid m.
   c. Trapezius m.

Muscles of the Pectoral Girdle and Trunk

See Figures 8.8–8.11 in your lab atlas.

1. Muscles of the thorax
   a. Pectoralis minor m.
   b. External intercostals m.
   c. Internal intercostals m.
   d. Diaphragm m.
2. Muscles of the abdominal wall
   a. Rectus abdominis m.
   b. External obliques m.
   c. Internal obliques m.
   d. Serratus anterior m.
   e. Transversus abdominis m.
3. Muscles of the posterior thorax
   a. Erector spinae m.
      (1) Iliocostalis m.
      (2) Longissimus m.
      (3) Spinalis m.

Muscles of the Upper Limb

See Figures 8.9–8.15 in your lab atlas.

1. Muscles of the shoulder
   a. Deltoid m.
   b. Pectoralis major m.
   c. Latissimus dorsi m.
2. Rotator cuff muscles
   a. Infraspinatus m.
   b. Subscapularis m.
   c. Supraspinatus m.
   d. Teres minor m.
3. Muscles of the arm
   a. Biceps brachii m.
   b. Triceps brachii m.
   c. Brachialis m.
   d. Brachioradialis m.
4. Muscles of the forearm
   a. Pronator teres m.
   b. Flexor carpi radialis m.
   c. Flexor carpi ulnaris m.
   d. Extensor carpi radialis longus m.
   e. Extensor digitorum m.
   f. Extensor carpi ulnaris m.

Muscles of the Pelvic Girdle and Lower Limb

See Figures 8.21–8.27 in your lab atlas.

1. Muscles of the pelvic girdle
   a. Iliopsoas m. (Iliacus m. and Psoas major m.)
   b. Gluteus maximus m.
   c. Gluteus medius m.

2. Muscles of the thigh
   a. Sartorius m.
   b. Tensor fasciae latae m.
   c. Quadriceps femoris group
      (1) Rectus femoris m.
      (2) Vastus medialis m.
      (3) Vastus intermedius m.
      (4) Vastus lateralis m.
   d. Adductor group
      (1) Adductor magnus m.
      (2) Adductor longus m.
      (3) Adductor brevis m.
   e. Hamstring muscles
      (1) Biceps femoris m.
      (2) Semimembranosus m.
      (3) Semitendinosus m.

3. Leg muscles
   a. Gastrocnemius m.
   b. Soleus m.
   c. Flexor digitorum longus m.
   d. Fibularis (peroneus) longus m.
   e. Fibularis (peroneus) brevis m.
   f. Tibialis posterior m.
   g. Tibialis anterior m.
   h. Extensor digitorum longus m.

> ➤ NOTE: YOUR INSTRUCTOR MAY WISH TO OMIT CERTAIN MUSCLES INCLUDED ABOVE, OR ADD MUSCLES NOT INCLUDED IN THESE LISTS. LIST ANY ADDITIONAL STRUCTURES BELOW:

Model Inventory

It might seem odd to use a model inventory for the muscular system; however, you will see that certain muscles will be visible on some models and not on others. It therefore becomes handy to have your model inventory to direct you to the proper model.

MODEL/DIAGRAM	STRUCTURES IDENTIFIED

EXERCISE 2: Muscle Actions

To fully understand the actions of a muscle, it is first necessary to discuss the muscle's origin and insertion. A muscle's *origin* is the part from which it originates, which is often immovable, and the muscle typically *inserts* into the part that it moves. For example, the sternocleidomastoid muscle originates on the sternum and clavicle, and crosses superomedially to the mastoid process of the skull, where it inserts.

Once you determine a muscle's origin and insertion, figuring out its actions becomes easy. Let's examine the sternocleidomastoid muscle again. It inserts into the head, so we know that it will move the head. Given its superomedial orientation, we can see that it will flex the neck at the cervical vertebrae and medially and laterally rotate the head at the atlantoaxial joint. Now wasn't that easy?

Building Muscles

In this exercise we will use small skeletons and modeling clay to build specific muscle groups. This may sound easy, but there is a catch: You must determine the actions of the muscle by looking only at the origin and insertion of each muscle that you build.

Procedure

1. Obtain a small skeleton and five colors of modeling clay.

2. Build the indicated muscles, using a different color of clay for each muscle. As you build, pay careful attention to the origin and insertion of each muscle.

3. Determine the primary actions for each muscle that you have built by looking *only* at the origin and insertion (no peeking at the Pre-Lab Exercises!).

a. Group 1:

MUSCLE	ACTIONS
Biceps femoris m.	
Semitendinosus m.	
Semimembranosus m.	
Gracilis m.	
Adductor group mm.	

b. Group 2:

MUSCLE	ACTIONS
Biceps brachii m.	
Triceps brachii m.	
Brachioradialis m.	
Deltoid m.	
Latissimus dorsi m.	

c. Group 3:

MUSCLE	ACTIONS
Pectoralis major m.	
Pectoralis minor m.	
Trapezius m.	
Masseter m.	
Temporalis m.	

3 EXERCISE 3: Demonstrating Muscles' Actions

This exercise should look familiar, as you did a similar exercise in Unit 9 (Articulations). As before, you will be performing various activities to demonstrate which joints are moving with each activity. In this exercise, however, rather than focusing on the motions occurring at each joint, you will be determining *which muscles* are causing the motions at each joint. Following is an example:

Example: Walking in Place

Part 1

First, list all motions occurring at each joint (the joints are listed with their common names, and the lists are certainly *not* all inclusive):

- Vertebral column and neck: extension
- Shoulder: flexion/extension
- Elbow: flexion
- Hip: flexion/extension, abduction/adduction
- Knee: flexion/extension
- Ankle: dorisflexion/plantarflexion, inversion/eversion

Part 2

Now refer to the chart in the Pre-Lab Exercises and the charts in Chapter 8 of your lab atlas to determine which muscles are causing each joint to move. Please note that these lists are far from complete. With nearly 700 muscles in the human body, a complete list could take all day!

- Vertebral column and neck
  - Extension: erector spinae and trapezius mm.
- Shoulder
  - Flexion: pectoralis major m.
  - Extension: latissimus dorsi m.
- Elbow
  - Flexion: biceps brachii, brachialis, and brachioradialis mm.
- Hip
  - Flexion: iliopsoas, sartorius mm.
  - Extension: biceps femoris, semitendinosus, semimembranosus, and gluteus maximus mm.
  - Abduction: tensor fasciae latae, sartorius, gluteus medius mm.
  - Adduction: gracilis, adductors mm.
- Knee
  - Flexion: biceps femoris, semitendinosus, semimembranosus, gastrocnemius mm.
  - Extension: rectus femoris, vastus medialis, vastus intermedius, vastus lateralis mm.

- Ankle
  - Dorsiflexion: tibialis anterior m.
  - Plantarflexion: gastrocnemius, soleus mm.
  - Inversion: tibialis anterior, tibialis posterior mm.
  - Eversion: fibularis brevis and fibularis longus mm.

Okay, so now you're thinking, *Wow, that's a lot of work!* Following are a few hints that will make this task less daunting:

- You will notice that the activities below are the same as the activities from Exercise 4 in Unit 9. To save yourself some work, reference Exercise 4 in Unit 9 to complete Part 1 of this exercise.

- As before, perform the indicated activity. You will miss some things if you don't!

- Unless your instructor asks you to do otherwise, try to keep it simple. List only the main muscles for each action. For example, when listing the muscles to flex the hip, we could easily include 10 different muscles. To make it simpler, list only those muscles we discussed that flex the hip as their *primary action* (as in the example on page 84).

Ready to give it a try?

Climbing the Stairs
Part 1:

Part 2:

Doing Jumping Jacks
Part 1:

Part 2:

Answering the Telephone
Part 1:

Part 2:

Jumping Rope

Part 1:

Part 2:

✔ Check Your Understanding

1. A condition called "drop foot," caused by certain neuro-muscular diseases, results in a patient's inability to dorsi-flex the foot. Which muscle(s) do you think is/are involved in this condition? Explain.

2. To compensate for drop foot, a patient adopts a gait pattern called a "steppage gait." What do you think that a steppage gait would look like? Which muscles would have to compensate to create this gait pattern?

3. One of the most commonly performed orthopedic surgeries is rotator cuff repair. Often this repair involves removal and reattachment of the deltoid muscle at its insertion. What functional impairment would be present following this procedure? What exercises could a physical therapist use to recondition the deltoid muscle?

4. Radical mastectomy is a type of surgery for breast cancer that involves removing all of the breast tissue, lymph nodes, and the pectoralis major muscle. What functional impairment would be present following this surgery? Which other muscles might a physical therapist train to enable the patient to regain some of this lost function?

12. Nervous Tissue

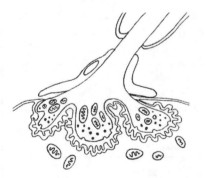

■ **OBJECTIVES** _____

Once you have completed this unit, you should be able to:

1. Describe the microanatomy of nervous tissue and the parts of a synapse.

2. Identify structures of the neuron and neuroglial cells on anatomical models and microscope slides.

■ **MATERIALS** _____

● Anatomical models: neuron, synapse, and axon end bulb

● Microscope slides: nervous tissue, nerve with myelin sheath, and the retina

● Modeling clay in four colors

● Colored pencils

PRE-LAB EXERCISES

Prior to coming to lab, complete the following exercises, referencing Chapters 3 and 9 in your lab atlas.

1 PRE-LAB EXERCISE 1: Nervous Tissue Microanatomy _____

Label and color-code Figure 12.1 using the terms from Exercise 1 (please note that all structures may not be visible on the diagrams). See Figures 9.2 and 9.3 in your lab atlas for reference.

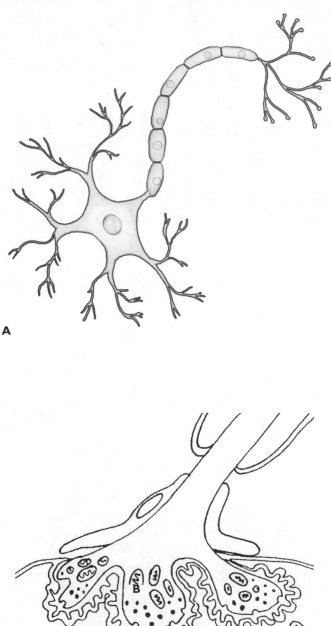

A

FIGURE 12.1
(A) Myelinated neuron with neuroglial cells;
(B) Axon end bulb with synapse.

2 PRE-LAB EXERCISE 2: Neuron Anatomy and Neuroglial Cells _____

Nervous tissue consists of two types of cells: (1) *neurons*, which generate and transmit nerve impulses, and (2) supporting cells called *neuroglial cells*. Fill in the following chart with the functions of the different neuroglial cells in the central and peripheral nervous systems and the parts of the neuron.

STRUCTURE	DEFINITION/FUNCTION
Parts of the Neuron:	
Axon	
Dendrite	
Soma (cell body)	
Nissl body	
Neuroglial cells:	
Oligodendrocyte	
Schwann cell	
Satellite cell	
Astrocyte	
Microglial cell	
Ependymal cell	
Other structures:	
Myelin sheath	
Node of Ranvier	
Synaptic vesicle	

EXERCISES

The nervous system is one of the major homeostatic systems in the body. It regulates cellular activities rapidly by sending nerve impulses down specialized cells called *neurons*. Neurons are supported by smaller cells called *neuroglial cells*, and together these two cell types make up *nervous tissue*.

The following exercises ask you to examine the structures of nervous tissue. Before proceeding, you may wish to review Unit 5, Histology.

EXERCISE 1:
Identification of Nervous Microanatomy ____

Within nervous tissue, there are two primary types of cells: *neurons* and *neuroglial cells*. Neurons are large cells that transmit and generate messages in the form of nerve impulses. They vary widely in size and structure, but most have the following features in common:

1. *Cell body*: The biosynthetic center of the neuron, the cell body contains the nucleus and much of the organelles. Its cytoskeleton consists of densely packed microtubules and *neurofibrils*, which compartmentalizes the rough endoplasmic reticulum into dark-staining structures called *Nissl bodies*. Cell bodies are often found in clusters. In the central nervous system, clusters of cell bodies are called *nuclei*, and in the peripheral nervous system, such clusters are called *ganglia*.

2. *Dendrites*: Neurons have one or more *dendrites*, which are processes that receive messages from sensory receptors or other neurons and transmit that information to the cell body.

3. *Axon*: A single axon exits the cell body to transmit messages to other neurons, muscles, and glands. It begins at the side of the cell body that contains the *axon hillock* and terminates in an *axon end bulb* (terminal button). Recall from Unit 10 that the axon end bulb contains synaptic vesicles with neurotransmitters that send messages to the axon's target. When the axon is communicating with another neuron, the location where they meet is called a *synapse*. The synapse has three parts:

   a. *Presynaptic neuron*: the neuron that is sending the message. It contains neurotransmitters packaged in synaptic vesicles in its axon end bulb.

   b. *Synaptic cleft*: the small space through which the neurotransmitters diffuse when they are released from the synaptic vesicles.

   c. *Postsynaptic neuron*: the neuron that receives the message. On its surface it has receptors for neurotransmitters. When the neurotransmitters bind, it

causes a change in the membrane potential of the postsynaptic neuron, either triggering or inhibiting a nerve impulse.

Neuroglial cells are much smaller than neurons, and they outnumber neurons about 50 to 1—no small feat considering that the nervous system contains about a trillion neurons! Neuroglial cells play a variety of roles, including anchoring neurons and blood vessels, circulating cerebrospinal fluid, and forming a structure called the *myelin sheath*. The myelin sheath, which covers the axons of certain neurons, is actually the cell membrane of Schwann cells (in the peripheral nervous system) and oligodendrocytes (in the central nervous system).

The myelin sheath functions to protect and insulate the axons and speed up conduction of nerve impulses. Because the sheath is made up of individual neuroglial cells, there are small gaps between the cells where the cell membrane of the axon is exposed. These gaps are called *nodes of Ranvier*, and the myelin-covered segments between the nodes are called *internodes*.

Identify the following parts of the neuron and neuroglial cells on models and diagrams, using your textbook and Figures 9.2 and 9.3 in your lab atlas as a guide. Refer to Pre-Lab Exercise 1 to review the locations of each of these structures.

1. Neuron
   a. Axon
   b. Dendrite(s)
   c. Cell body (soma)
   d. Nissl bodies
   e. Neurofibrils

2. Neuroglial cells
   a. Oligodendrocytes
   b. Schwann cells
   c. Satellite cells
   d. Microglial cells
   e. Astrocytes
   f. Ependymal cells

3. Other structures
   a. Myelin sheath
   b. Node of Ranvier
   c. Axon hillock
   d. Nucleus
   e. Ganglion

4. Parts of the synapse
   a. Presynaptic neuron
   b. Axon end bulb

c. Synaptic vesicle

d. Synaptic cleft

e. Postsynaptic neuron

f. Neurotransmitter receptors

Model Inventory

As you view the anatomical models in lab, list them on the inventory below, and detail which of the above structures you are able to locate on each model.

MODEL / DIAGRAM	STRUCTURES IDENTIFIED

Building a Myelin Sheath

In this exercise you will demonstrate the difference between the methods by which Schwann cells and oligodendrocytes myelinate the axons of neurons in the peripheral nervous system and central nervous system, respectively. Both types of cells create the myelin sheath by wrapping themselves around the axon repeatedly (as many as 100 times). Recall, however, that Schwann cells can myelinate only one axon, while oligodendrocytes send out "arms" to myelinate several axons in its vicinity.

Procedure

1. Obtain four colors of modeling clay (blue, green, yellow, and red, if available).

2. Use the following color code to build three central nervous system (CNS) axons, one peripheral nervous system (PNS) axon, one oligodendrocyte, and two Schwann cells.

   ● CNS axons: blue

   ● PNS axon: green

   ● Oligodendrocyte: yellow

   ● Schwann cells: red

3. Build your oligodendrocyte so it reaches out to myelinate the three CNS axons.

4. Build your Schwann cells so that they myelinate the single PNS axon, making sure to leave a gap for the node of Ranvier.

✔ Check Your Understanding

1. Multiple sclerosis is a *demyelinating* disease, in which the patient's immune system attacks and destroys the myelin sheath in the central nervous system. What symptoms would you expect from such a disease? Why? Would Schwann cells or oligodendrocytes be affected?

2. Would you expect to see synaptic vesicles in dendrites? Why or why not? (*Hint*: Consider the functions of axons versus dendrites.)

3. Neurons are amitotic, meaning that, after a certain stage, they do not divide further. Cancerous cells are characterized by having a rapid rate of mitosis. Considering this, of which cell types (neurons or neuroglia) must brain tumors be composed? Why?

2 EXERCISE 2: Histology of Nervous Tissue _____

Neurons may be classified on the basis of the number of processes extending from the cell body:

1. *Multipolar neurons*: These neurons have three or more processes—specifically, one axon and two or more dendrites. Although technically a multipolar neuron may have only two dendrites, most have hundreds or more, and look like a highly branched tree. More than 99% of all neurons are multipolar, and these are what are typically present on most prepared slides of nervous tissue.

2. *Bipolar neurons*: Much rarer are the bipolar neurons, which have one axon and one dendrite. Bipolar neurons tend to be found in special sense organs such as the olfactory epithelium and the retina of the eye.

3. *Unipolar neurons*: These neurons, found in the skin for sensations of touch and pain, have only one process extending from the cell body. A short distance from the cell body, it branches like a "T," with one process coming from the sensory receptor and the other going toward the spinal cord. A unique feature of this type of neuron is that the entire process is considered an axon, and no dendrites enter the cell body.

In this exercise you will examine three microscope slides:

1. *Nervous tissue*: This is perhaps the easiest slide to identify, as it should contain many multipolar neurons. Surrounding the large neurons you will see many much smaller cells (they may look like purple dots). These are the neuroglial cells. As you scan the tissue on low power, find a well-stained neuron and move the objective lens to high power. Look for Nissl bodies and try to identify the single axon and the axon hillock. (This may be difficult, so don't get discouraged if all of the processes look similar.)

2. *Retina*: This tissue shows a good example of bipolar neurons. As you focus in on the retina, compare the structure of these neurons to those from the slide of nervous tissue.

3. *Myelin sheath*: A longitudinal section of axons stained with a stain specific for the components of myelin (often osmium) is useful for seeing the sheath itself and also the nodes of Ranvier. To see the nodes, you will likely need to put the objective lens to high power.

Examine the following prepared histological sections on microscope slides. Use colored pencils to draw what you see, and label your drawing with the terms indicated.

1. **Nervous Tissue** (See Figures 3.41 and 3.49–3.51 in your lab atlas)
   a. Cell body
   b Axon
   c. Dendrites
   d. Nissl bodies
   e. Neurofibrils

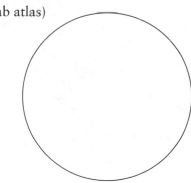

2. **Retina** (See Figure 11.6 in your lab atlas)
   a. Bipolar neurons
   b. Axons
   c. Dendrites

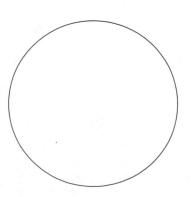

3. Myelin Sheath (See Figure 3.45 in your lab atlas)
   a. Axon with myelin sheath
   b. Node of Ranvier

13. Central Nervous System

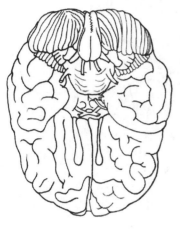

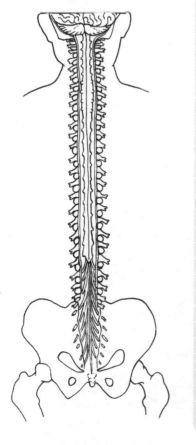

■ **OBJECTIVES**

Once you have completed this unit, you should be able to:

1. Describe the gross anatomy of the brain.

2. Identify structures of the brain on anatomical models and preserved or fresh specimens.

3. Identify structures of the spinal cord on anatomical models and preserved or fresh specimens.

■ **MATERIALS**

● Anatomical models: brain, ventricles of the brain, spinal cord, and vertebral column

● Preserved or fresh sheep brains

● Dissection equipment

PRE-LAB EXERCISES

Prior to coming to lab, complete the following exercises, using Chapter 9 in your atlas for reference.

1 | PRE-LAB EXERCISE 1: Anatomy of the Brain_____

Label and color-code Figure 13.1 using the terms from Exercise 1 (please note that all terms may not be visible on the diagrams). See Figures 9.5 and 9.7–9.31 in your lab atlas for reference.

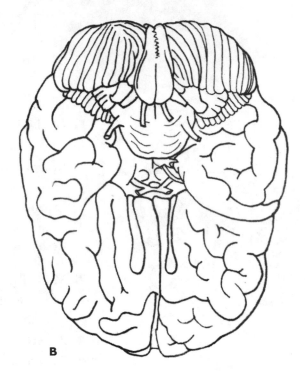

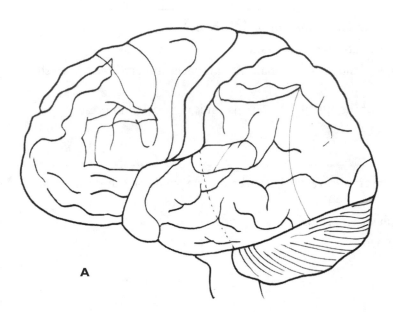

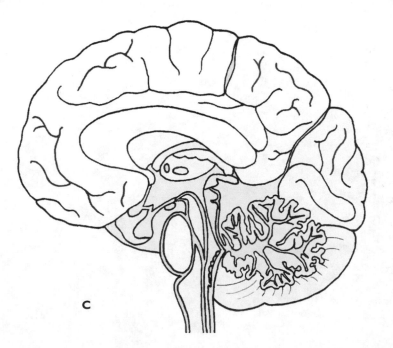

FIGURE 13.1
Views of the brain from (A) lateral side, (B) inferior side, (C) sagittal section.

2 PRE-LAB EXERCISE 2:
Anatomy of the Spinal Cord _____

Label and color-code Figure 13.2, using the terms from Exercise 2. See Figures 9.1 and 9.4–9.6 in your atlas for reference.

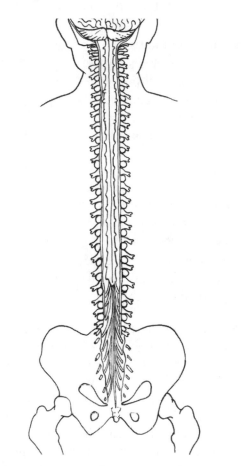

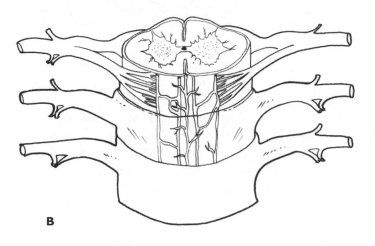

FIGURE 13.2
Diagram of (A) spinal cord in vertebral column, (B) transverse section of spinal cord.

EXERCISES

The central nervous system (CNS) consists of two major parts: the brain and the spinal cord. Each of these will be explored in the three exercises in this chapter.

1 EXERCISE 1:
Anatomy of the Brain _____

The brain is a complex organ that is divided generally into four regions:

1. the cerebral hemispheres (called collectively the *cerebrum*),
2. the diencephalon,
3. the brainstem, and
4. the cerebellum.

Within the brain are hollow chambers called *ventricles*, which are filled with *cerebrospinal fluid* (CSF), a fluid similar to plasma without the proteins that bathes the brain and spinal cord. The largest of the ventricles, the *lateral ventricles*, are located in each cerebral hemisphere. When viewed from the anterior side, they resemble rams' horns. The smaller *third ventricle* is housed within the diencephalon, and is continuous with the *fourth ventricle*, found in the brainstem, by a small canal called the *cerebral aqueduct*. The fourth ventricle is continuous with the central canal of the spinal cord.

Recall from Unit 12 that the ventricles are lined by neuroglial cells called ependymal cells, whose cilia beat to circulate CSF. Something interesting to note about CSF is that one of its main functions is to reduce brain weight (the brain is buoyant in the CSF). Without CSF, your brain would literally crush itself under its own weight!

The brain is surrounded by a set of three membranes collectively called the *meninges*.

1. *Dura mater*: The outermost meninx layer, the dura mater is a thick, leathery, double-layered membrane. The superficial layer is fused to the skull; the other is continuous with the dura mater of the spinal cord. Over most of the brain, the two layers of the dura are fused. In three regions, however, the inner layer of the dura separates from the outer layer and dives into the brain:
   a. the *falx cerebri*, which forms a partition between the right and left cerebral hemispheres
   b. the *falx cerebelli*, which separates the cerebrum from the cerebellum
   c. the *tentorium cerebelli*, which separates the two cerebellar hemispheres.
   As the inner layer dives in to form the falx cerebri, a space between the two layers is created, called the *superior sagittal sinus*. The superior sagittal sinus is part of a series of *dural sinuses* that drain venous blood from the brain.
2. *Arachnoid mater*: The middle meninx, the arachnoid mater is separated from the dura by a CSF-filled space called the *subarachnoid space*. This space contains projections of the arachnoid mater called the *arachnoid villi*, which allow the CSF to reenter the blood via the dural sinuses.
3. *Pia mater*: The thinnest, innermost meninx is the pia mater. Rich with blood vessels, the pia mater clings to the surface of the cerebral hemispheres.

The *cerebral hemispheres* (together called the *cerebrum*) are the superiormost portion of the brain. They have a surface consisting of elevated ridges called *gyri* and shallow grooves called *sulci*; deep grooves, called *fissures*, separate major regions of the cerebral hemispheres. The *longitudinal fissure* separates the right and left hemispheres.

The cerebrum consists of five lobes: the *frontal*, *parietal*, *temporal*, *occipital*, and deep *insula* lobes (remember this last one by the mnemonic "the *insula* is *insula*ted"). The cerebral hemispheres are responsible for cognitive functions including learning and language, conscious interpretation of sensory information, conscious planning of movement, and personality.

The cell bodies of the hemispheres lie in the outer 2 millimeters in a region called the *cerebral cortex*. It consists of unmyelinated fibers and cell bodies, and is called *gray matter* because it is gray in color. Most of the remainder of the cerebral hemispheres is white matter, consisting of myelinated axons that communicate with other regions of the brain. The largest tract of cerebral white matter, the *corpus callosum*, connects the right and left cerebral hemispheres.

Deep to the cerebral hemispheres is a structure called the diencephalon, composed of three major regions: the *thalamus*, the *hypothalamus*, and the *epithalamus*.

1. The *thalamus*, which makes up 80% of the diencephalon, is a major integration and relay center that edits and sorts information going into and out of the cerebrum.
2. The *hypothalamus*, located on the anterior and inferior aspect of the diencephalon, is a deceptively small structure that carries out many of the body's homeostatic functions, including helping to regulate the endocrine system, monitoring the sleep–wake cycle, controlling thirst, hunger, and body temperature, and helping to monitor the autonomic nervous system (I guess it's true—good things do come in small packages!). The *pituitary gland* is connected to the hypothalamus by a stalk called the *infundibulum*.
3. The *epithalamus* contains a gland called the *pineal gland*, which secretes the hormone *melatonin*, important in regulating the sleep/wake cycle.

The *brainstem*, the third major portion of the brain, controls the automatic functions of the body, such as the rhythm for breathing, the heart rate, blood pressure, and certain reflexes. The most superior portion of the brainstem, the *midbrain*, sits above the *pons*, the rounded segment that bulges anteriorly. The last segment of the brainstem, the *medulla oblongata*, merges with the spinal cord.

The *cerebellum*, the fourth major component of the brain, consists of two highly convoluted lobes connected by a piece called the *vermis*. Like the cerebral hemispheres, the cerebellum has an outer *cerebellar cortex*, composed of gray matter, and inner white matter often called the *arbor vitae* because of its resemblance to the branches of a tree. The cerebellum works to coordinate motor activities, receiving input from proprioceptors, the inner ear, the optic nerves, and the cerebrum.

Identify the following structures of the brain on anatomical models and diagrams. See Figures 9.5 and 9.7–9.21 in your atlas for reference.

Structures of the Brain

1. Cerebral hemispheres
2. Corpus callosum (cerebral white matter)
3. Cerebral cortex (gray matter)
4. Basal nuclei or ganglia (these will only be visible on certain transverse sections of the brain)

5. Lobes of the cerebrum
   a. Frontal lobe
   b. Parietal lobe
   c. Occipital lobe
   d. Temporal lobe
   e. Insula lobe

6. Fissures
   a. Longitudinal fissure
   b. Transverse fissure

7. Sulci
   a. Central sulcus
      (1) Pre-central gyrus
      (2) Post-central gyrus
   b. Lateral sulcus
   c. Parieto-occipital sulcus

8. Diencephalon
   a. Thalamus
   b. Hypothalamus
      (1) Infundibulum
      (2) Pituitary gland
   c. Epithalamus
   d. Pineal gland

9. Brainstem
   a. Midbrain
   b. Pons
   c. Mcdulla oblongata

10. Cerebellum
    a. Vermis
    b. Arbor vitae

11. Brain coverings
    a. Dura mater
       (1) Subdural space
       (2) Falx cerebri
       (3) Falx cerebelli
    b. Arachnoid mater
       (1) Subarachnoid space
    c. Pia mater

12. Ventricles
    a. Lateral ventricles
    b. Third ventricle
    c. Fourth ventricle
    d. Cerebral aqueduct

13. Vascular structures
    a. Dural sinuses
       (1) Superior sagittal sinus
       (2) Transverse sinus
       (3) Cavernous sinus
    b. Choroid plexus
    c. Arachnoid villi

Model Inventory

As you examine the anatomical models and diagrams in lab, list them on the inventory below, and detail which structures you are able to locate on each model.

MODEL/DIAGRAM	STRUCTURES IDENTIFIED

✔ Check Your Understanding

1. A condition that may result from traumatic brain injuries is *cerebral edema* (swelling or fluid on the brain). Do you think that cerebral edema would be more damaging to an adult or to an infant? Why? (*Hint*: Think about the structure of the skull in the adult versus an infant.)

2. Predict the effects of injuries to the following areas (you may want to use your text for this one):

a. Frontal lobes:

b. Brainstem:

c. Cerebellum:

d. Occipital lobes:

e. Basal nuclei:

3. Which of the above injuries do you think would be the most damaging to survival? Why?

2 EXERCISE 2: The Spinal Cord

As the medulla oblongata passes through the foramen magnum of the occipital bone, it becomes the *spinal cord*. The spinal cord extends to approximately the first or second lumbar vertebra, where it tapers to the *conus medullaris*. From the conus medullaris arises a tuft of nerve roots called the *cauda equina* (meaning "horse's tail"), which continues down the vertebral column to the sacrum, exiting out of the appropriate vertebrae to become spinal nerves.

The meninges that cover the brain are continuous with the spinal cord. An important difference, however, is that the dura mater of the spinal cord consists of only one layer rather than two and it does not attach to the vertebral column. This gives rise to an *epidural space* not seen around the brain.

Another difference is the presence of small extensions of pia mater called *denticulate ligaments*, which secure the spinal cord to the vertebral column. The dura and the arachnoid mater both extend beyond the conus medullaris, down to about the level of S2. The pia mater travels even farther, forming a long, fibrous extension called the *filum terminale*, which eventually attaches to the coccyx. Like the meninges of the brain, those of the spinal cord are filled with cerebrospinal fluid, which bathes and cushions the spinal cord.

Internally, the spinal cord consists of a butterfly-shaped inner core of gray matter, which surrounds the cerebrospinal fluid-filled *central canal*. The gray matter is divided into regions, or horns. Anteriorly are the *anterior* (ventral) *horns*, which contain the cell bodies of motor neurons. The anterior horn gives off axons to form the *ventral root*—a collection of motor fibers that eventually will become part of a spinal nerve. On the posterior side are the *posterior* (dorsal) *horns*, which hold association neurons that receive fibers from the *dorsal root* (also part of the spinal nerve). The dorsal root contains sensory axons, whose cell bodies are located in a ganglion called the *dorsal root ganglion*. The dorsal root ganglion (and therefore the dorsal root) is easily identified as a large, swollen knob just outside the spinal cord. The anterior and posterior horns are distinguishable by their shapes: The anterior horns are broader and flatter on the ends, whereas the posterior horns are more tapered, and they extend farther out toward the edge. In the thoracic and lumbar regions of the spinal cord on the lateral edges there are additional horns, the *lateral horns*, which contain the cell bodies of autonomic (specifically, sympathetic) neurons.

Outside the gray matter is the spinal white matter, which can be divided into three regions or columns: the *anterior, posterior,* and *lateral funiculi*. Each funiculus contains myelinated axons that can be grouped into *tracts*. Tracts contain axons that have the same beginning and end points, and the same general function. Ascending tracts carry sensory information from sensory neurons to the brain, and descending tracts carry motor information from the brain to motor neurons.

Structures of the Spinal Cord

Identify the following structures of the spinal cord on anatomical models and diagrams. See Figures 9.4–9.6, 9.21, and 9.22 in your lab atlas and Pre-Lab Exercise 2 for reference.

1. Meninges
   a. Dura mater
      (1) Epidural space
   b. Arachnoid mater
   c. Pia mater
      (1) Denticulate ligaments
2. Spinal gray matter
   a. Anterior horn
   b. Lateral horn
   c. Posterior (dorsal) horn

3. Spinal white matter
   a. Anterior funiculus
   b. Lateral funiculus
   c. Posterior funiculus

4. Spinal nerve roots:
   a. Dorsal root
      (1) Dorsal root ganglion
   b. Ventral root

5. Conus medullaris

6. Filum terminale

7. Cauda equina

8. Cervical enlargement

9. Lumbar enlargement

10. Central canal

11. Anterior median fissure

12. Posterior median sulcus

Model Inventory

As you examine the anatomical models and diagrams in lab, list them on the inventory below, and detail which structures you are able to locate on each model.

Model / Diagram	Structures Identified

✔ Check Your Understanding

1. A common way to deliver anesthesia for surgery and childbirth is to inject the anesthetic agent into the epidural space (hence the term *epidural anesthesia*). A possible complication of this procedure is a tear in the dura mater, and the resulting leakage of cerebrospinal fluid out of the central nervous system. Why would a loss of cerebrospinal fluid be problematic? What symptoms do you predict with this condition? (*Hint*: Think about the function of cerebrospinal fluid.)

2. A similar procedure is the *lumbar puncture*, in which cerebrospinal fluid surrounding the meninges is withdrawn. This procedure is generally performed between L3 and L5. Why do you think the fluid is withdrawn here rather than from higher up in the vertebral column (e.g., the cervical vertebrae)? Often the sampled CSF is tested for viruses if viral encephalitis is suspected. Because encephalitis is an infection of the brain, why would CSF sampled from the spinal column give you information about the condition of the brain?

3 EXERCISE 3: Brain Dissection

This exercise will allow you to examine structures that are often difficult to perceive on anatomical models, such as the ventricles, by dissecting preserved sheep brains. You will note that certain structures, such as the frontal lobes of the cerebral hemispheres, are proportionally smaller in the sheep than in the human brain.

Procedure

> NOTE: GOGGLES AND GLOVES ARE REQUIRED.

1. If the brain is still encased in the skull, you have your work cut out for you. The best way to approach extracting it from the skull is to take a hammer and chisel, and gently (at least as gently as one can with a hammer and chisel) remove it piece by piece.

2. As you gently hack away at the skull, you will note a thick membrane holding the skull in place. This is the dura mater, and it can make removal of the skull somewhat difficult. Ideally, you would like to preserve the dura, but you may end up cutting through it as you remove the brain.

3. Once you have removed most of the skull, gently lift out the brain. (If you're careful, you may be able to get the brain out with the pituitary gland still attached.) You may have to loosen the remaining attachments of the dura with your finger.

4. Once the brain is out, note the thick part of the dura covering the longitudinal fissure. If you cut through this with scissors, you will enter the superior sagittal sinus.

5. Next, remove the dura to reveal the thin membrane on top of the brain. This is the arachnoid mater.

6. Remove an area of the arachnoid mater to see the shiny inner membrane—the pia mater—directly touching the surface of the brain. Note that the pia mater follows the convolutions of the gyri and sulci.

7. Examine the surface anatomy of both the superior and the inferior surfaces of the sheep brain. Below, list all structures from Exercise 1 that you are able to identify. Refer to Figures 9.23–9.26 in your lab atlas for help.

8. Note the size of the olfactory bulbs. Are they larger or smaller than those that you observed in the human brain? Why do you think this is so?

9. Spread the two cerebral hemispheres and identify the corpus callosum.

10. To view the internal anatomy, make a cut down the mid-sagittal plane of the brain, separating the two cerebral hemispheres through the corpus callosum, and sectioning the diencephalon, cerebellum, and brainstem.

11. Examine the brain's internal anatomy, and list below all structures you are able to identify. See Figures 9.27–9.29 in your lab atlas for reference.

12. As you examine the internal anatomy, stick your finger in the lateral ventricle. You will see (or feel) that it is much larger than it appears.

13. Section one of the halves of the brain in the frontal plane, approximately along the central sulcus. Note the outer cerebral cortex (the gray matter) and the inner white matter. You also can see the lateral and third ventricles from this view.

14. If the brain on which you are working contained a segment of the spinal cord, move to it now, and identify the single-layered dura surrounding it. Peel it back to see the arachnoid mater.

15. Section the cord in the transverse plane. Note the butterfly-shaped inner gray matter and the outer white matter. Below, list the structures that you can identify. See Pre-Lab Exercise 2 for reference.

14. Peripheral and Autonomic Nervous System

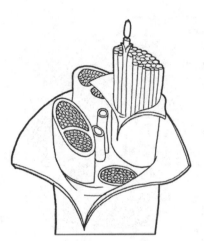

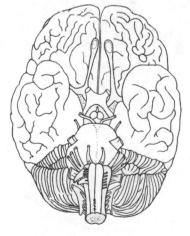

■ OBJECTIVES

Once you have completed this unit, you should be able to:

1. Identify structures of a peripheral nerve on anatomical models or preserved specimens.

2. Identify, describe, and demonstrate functions of cranial nerves.

3. Identify spinal nerves and plexuses on anatomical models.

4. Describe a simple spinal reflex arc and test the effects of mental concentration on the patellar tendon (knee-jerk) reflex.

5. Describe the effects of the two branches of the autonomic nervous system on the body systems.

■ MATERIALS

● Anatomical models: peripheral nerve, spinal nerves, spinal cord cross sections, brains

● Colored pencils

● For the cranial nerves experiment:

    Pen lights

    Snellen eye chart

    Tuning forks

    Samples of substances for smell identification (peppermint, coffee, orange)

    Tasting papers (PTC, thiourea, sodium benzoate)

● Reflex hammers

● Stethoscopes

● Sphygmomanometers

PRE-LAB EXERCISES

Prior to coming to lab, complete the following exercises, using Chapters 3 and 9 in your atlas for reference.

1 PRE-LAB EXERCISE 1:
Anatomy of a Peripheral Nerve _____

Label and color-code the diagram in Figure 14.1, depicting a nerve, and using the terms from Exercise 1. Please note that all structures may not be visible on the diagram. See Figure 3.43 in your lab atlas for reference.

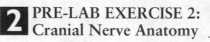

2 PRE-LAB EXERCISE 2:
Cranial Nerve Anatomy _____

Label and color-code the diagram of the brain in Figure 14.2, depicting the cranial nerves and using the terms from Exercise 2. See Figures 9.10 and 9.11 in your lab atlas for reference.

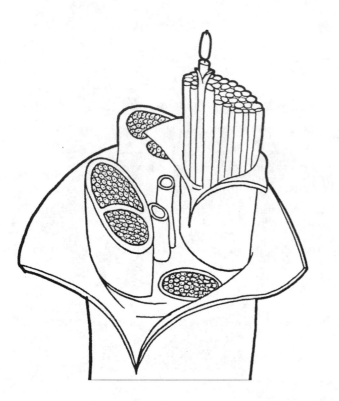

FIGURE 14.1
Diagram of a peripheral nerve.

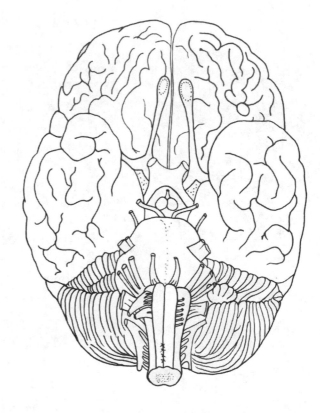

FIGURE 14.2
Diagram of brain, inferior view, with cranial nerves.

3 PRE-LAB EXERCISE 3:
Functions of Cranial Nerves

Cranial nerves, part of the peripheral nervous system, are those that originate from the brain and the brainstem. Fill in the chart with the following information regarding the 12 pairs of cranial nerves: functions, location (meaning from which part of the brain the nerve originates), and whether the nerve is motor, sensory, or mixed. *(Please note that no nerve is purely motor, as all motor nerves carry sensory proprioceptive fibers as well.)*

CRANIAL NERVE	FUNCTIONS	LOCATION	MOTOR, SENSORY, OR MIXED
CN I: Olfactory nerve			
CN II: Optic nerve			
CN III: Oculomotor nerve			
CN IV: Trochlear nerve			
CN V: Trigeminal nerve			
CN VI: Abducens nerve			
CN VII: Facial nerve			
CN VIII: Vestibulocochlear nerve			
CN IX: Glossopharyngeal nerve			
CN X: Vagus nerve			
CN XI: Accessory nerve			
CN XII: Hypoglossal nerve			

4 PRE-LAB EXERCISE 4: Spinal Nerves

Spinal nerves are those originating from the spinal cord. Each spinal nerve can be traced to a specific location on the spinal cord from which it originates, called the *spinal root*. Fill in the following chart with the spinal roots and structures innervated by each of the listed spinal nerves.

SPINAL NERVE	SPINAL ROOT	FUNCTIONS
Phrenic nerve		
Radial nerve		
Musculocutaneous nerve		
Ulnar nerve		
Median nerve		
Intercostal nerves		
Femoral nerve		
Sciatic nerve		
Tibial nerve		
Common peroneal (fibular) nerve		

5 PRE-LAB EXERCISE 5: Functions of the Autonomic Nervous System

In this unit we will cover a branch of the nervous system called the autonomic nervous system (ANS), which has two subdivisions: (1) the sympathetic nervous system (SNS), and (2) the parasympathetic nervous system (PSNS). Fill in the following chart with the general characteristics of each branch of the ANS, as well as the effects of each system on individual target organs.

CHARACTERISTIC	SNS	PSNS
Main function		
Location of nerve roots		
Neurotransmitter		
Effects on target organs and physiologic processes		
Heart		
Bronchioles (airway passages of the lungs)		
Blood vessels to abdominal organs and skin		
Blood vessels to skeletal muscle and cardiac muscle		
Digestive and urinary functions		
Metabolism		
Pupils		

EXERCISES

In the previous chapters we discussed the structure of neurons. We know that axons carry information from the cell body to the target, which can be a muscle, a gland, or another nerve. But axons, also called *nerve fibers,* do not travel to their respective destinations alone. They travel packaged together in a bundle collectively called a *nerve.* The peripheral nervous system (PNS) consists of nerves that originate from the brain and the brain stem, called *cranial nerves,* and nerves that originate from the spinal cord, called *spinal nerves.*

In the following exercises, you will examine the structure and functions of the peripheral nervous system, including a specialized branch of the PNS called the *autonomic nervous system.*

1 EXERCISE 1: Structure of a Peripheral Nerve _____

A nerve is an organ consisting of anywhere from a few nerve fibers to thousands of nerve fibers, as well as blood vessels, lymphatics, and connective tissue sheaths. Each individual nerve fiber is covered by its cell membrane, which may be covered by the myelin sheath (see Unit 12 for a review). On top of the cell membrane or myelin sheath is a connective tissue sheath called the *endoneurium.* Nerve fibers are grouped into larger bundles called *fascicles,* which are covered with another connective tissue sheath, the *perineurium.* Groups of fascicles, together with blood vessels and lymphatic vessels, are wrapped in the outermost connective tissue sheath, the *epineurium.*

Identify the following parts of the nerve on models and diagrams using your textbook and Figures 3.42–3.45 in your lab atlas as a guide. Refer to Pre-Lab Exercise 1 to review the anatomy of each of these structures.

Nerve Anatomy

1. Nerve
2. Fascicle
3. Axon (nerve fiber)
4. Blood vessels
5. Nerve sheaths
   a. Epineurium
   b. Perineurium
   c. Endoneurium
   d. Myelin sheath

Model Inventory

As you view the anatomical models in lab, list them on the inventory below, and detail which of the above structures you are able to locate on each model.

MODEL/DIAGRAM	STRUCTURES IDENTIFIED

2 EXERCISE 2: The Cranial Nerves _____

The 12 pairs of *cranial nerves* originate from the brain and brain stem. Each is given two names: (1) a sequential Roman numeral, and (2) a descriptive name. For example, cranial nerve III is the third cranial nerve to arise from the brain. It is also called the oculomotor nerve because one of its functions is to provide motor fibers to some of the muscles that move the eyeball. To remember the names and the order of the nerves, a helpful mnemonic is:

Oh (Olfactory)
Once (Optic)
One (Oculomotor)
Takes (Trochlear)
The (Trigeminal)
Anatomy (Abducens)
Final (Facial)
Very (Vestibulocochlear)
Good (Glossopharyngeal)
Vacations (Vagus)
Are (Accessory)
Happening (Hypoglossal)

You also can help yourself remember the olfactory and optic nerves by reminding yourself that you have one nose (CN I, the *olfactory* nerve) and two eyes (CN II, the *optic* nerve).

All cranial nerves innervate structures of the head and neck, but cranial nerve X, the vagus nerve, also leaves the head and neck and innervate other structures (the thoracic and abdominal viscera). Most cranial nerves carry both sensory and motor fibers, although three of them have purely sensory functions (which three were they?). Refer to Pre-Lab Exercise 3 to review the functions of each nerve.

Most of the pairs of cranial nerves remain separate throughout their courses. An exception is seen in the optic nerves, whose fibers emerge from the retina of the eye and meet at the *optic chiasma*. Here the nerves partially exchange fibers before diverging to form the optic tracts, which go to the thalamus before reaching the optic cortex of the occipital lobe.

Cranial Nerve Anatomy

Identify cranial nerves I–XII and the following structures on anatomical models and/or preserved specimens of the brain. For reference, see Figures 9.10 and 9.11 in your lab atlas, as well as Pre-Lab Exercise 2. List the nerves and the structures on the model inventory from Exercise 1.

1. Olfactory bulbs
2. Optic chiasma
3. Optic tract

Testing the Cranial Nerves

A component of every complete physical examination performed by health care professionals is the cranial nerve exam. In this exercise, you will put on your "doctor hat" and perform the same tests of the cranial nerves that would be done during a physical examination. Pair up with another student and take turns performing the following tests. For each test, first document your observations (in many cases, this will be "able to perform," or "unable to perform"). Then, state which cranial nerve(s) you have checked with each test. Keep in mind that some tests check more than one cranial nerve, some cranial nerves are tested more than once, and each nerve is tested in this exercise at least once.

Procedure

1. Have your partner perform the following actions individually (not all at once—your partner may find it difficult to smile and frown at the same time!): smile, frown, raise the eyebrows, puff the cheeks, open and close the jaw, clench the teeth, raise the shoulders.

Observations:

CN(s) tested:

2. Draw a large, imaginary "Z" in the air with your finger. Have your partner follow your finger with his/her eyes without moving his/her head. Repeat the procedure, this time drawing the letter "H" in the air with your finger.

Observations:

CN(s) tested:

3. Test pupillary response:
   a. Place your hand vertically against the bridge of your partner's nose. This forms a light shield to separate the right and left visual fields.

b. Shine the penlight indirectly into the left eye. Watch what happens to the pupil in the left eye.

c. Move the penlight away and watch what happens to the pupil.

d. Shine the light into the left eye again and watch what happens to the pupil in the *right* eye.

e. Repeat this process, shining the penlight into the right eye.

f. Fill in the chart below with your results.

ACTION	RESPONSE OF LEFT PUPIL	RESPONSE OF RIGHT PUPIL
Light shined into left eye		
Light removed from left eye		
Light shined into right eye		
Light removed from right eye		

CN(s) tested:

4. Have your partner focus on an object on the far side of the room (e.g., the blackboard or a chart) for 1 minute. Then have your partner switch his/her focus to an object in your hand (e.g., a pencil). Watch your partner's pupils carefully as the point of focus is changed from far to near.

Observations:

CN(s) tested:

5. Place your hand lightly on your partner's throat, and have him/her swallow. Feel for symmetrical movement of the larynx (throat).

Observations:

CN(s) tested:

6. Have your partner protrude his/her tongue. Check for abnormal deviation or movement (e.g., does the tongue move straight forward or does it move to one side?).

Observations:

CN(s) tested:

7. Examine your partner's throat with a penlight, and have him/her say, "Ahh." Check for symmetrical movement of the uvula and soft palate.

Observations:

CN(s) tested:

8. Test your partner's vision by having him/her stand 20 feet from a Snellen chart and read the chart, starting at the largest line and progressing to the smallest line he/she is able to see clearly. Record the ratio (e.g., 20/30) next to the smallest line your partner can read.

Observations:

CN(s) tested:

9. Hold a tuning fork by its handle and strike the tines with a rubber mallet (or, just tap it lightly on the lab table). Touch the stem to the top of your partner's head, along the midsagittal line. Ask your partner if he/she hears the vibration better in one ear or if he/she hears the sound equally well in both ears.

Observations:

CN(s) tested:

10. Have your partner stand with his/her eyes closed and arms at his/her sides for several seconds. Evaluate his/her ability to remain balanced.

Observations:

CN(s) tested:

11. Hold an unknown sample under your partner's nose, and have him/her identify the substance by its scent.

Observations:

CN(s) tested:

12. Evaluate your partner's ability to taste using tasting papers. Have your partner place a piece of PTC paper on his/her tongue and determine if he/she can taste it (the ability to taste PTC is genetically determined, about half of the population can taste it). If he/she cannot taste the PTC, try the thiourea paper instead (a word of warning: the thiourea tastes pretty bad). If your partner cannot taste either of these papers, try the sodium benzoate paper.

Observations:

CN(s) tested:

✔ **Check Your Understanding**

1. What conditions might produce abnormal results on a cranial nerve physical examination?

2. What happens to the pupil of the right eye when light is shined into the left eye (and vice versa)? Why? What could it indicate if the pupils did not respond equally? What if the pupils didn't respond at all?

3 EXERCISE 3: Spinal Nerves and Reflexes

There are 31 pairs of *spinal nerves* that form as the dorsal and ventral roots of the spinal cord fuse. Recall that the ventral roots carry motor fibers emerging from the spinal cord and the dorsal roots carry sensory fibers to the spinal cord. Because each spinal nerve carries both motor and sensory fibers, all spinal nerves are mixed nerves.

Shortly after the dorsal and ventral roots fuse to form the spinal nerve, it splits into three branches:

1. A *dorsal ramus*

2. A *ventral ramus*

3. A small *meningeal branch*.

The dorsal rami serve the skin, joints, and musculature of the posterior trunk, and the meningeal branches reenter the vertebral canal to innervate spinal structures. The larger ventral rami of the cervical, lumbar, and sacral nerves travel anteriorly to form four large *plexuses*, or networks, of nerves: the cervical, brachial, lumbar, and sacral plexuses. The ventral rami of the thoracic nerves do not form plexuses; they travel anteriorly as 11 separate pairs of intercostal nerves.

Spinal Nerve Anatomy

Identify the following plexuses and spinal nerves on anatomical models and diagrams. See Figures 9.1 and 9.21 in your atlas and your Pre-Lab Exercises for reference.

1. Cervical plexus
   a. Phrenic nerve

2. Brachial plexus
   a. Radial nerve
   b. Musculocutaneous nerve
   c. Median nerve
   d. Ulnar nerve

3. Thoracic (intercostal) nerves

4. Lumbar plexus
   a. Femoral nerve

5. Sacral plexus
   a. Sciatic nerve
      (1) Tibial nerve
      (2) Common peroneal (fibular) nerve

Spinal Reflexes

A *reflex* is an involuntary, predictable motor response to a stimulus. The pathway through which information travels, shown below, is called a *reflex arc*.

sensory receptor → sensory afferent nerve fibers →
central nervous system → motor efferent nerve fibers → muscle

The human body has many different reflex arcs, one of the least complicated of which is the *stretch reflex*. Stretch reflexes are important in maintaining posture and equilibrium and are initiated when a muscle is stretched.

The stretch is sensed by *muscle spindles*, specialized stretch receptors in the muscles, and this information is sent via afferent fibers to the central nervous system. The CNS then sends impulses down the motor efferent fibers, causing the muscle to shorten (contract), thereby countering the stretch.

You can demonstrate the stretch reflex easily: Sit down, with your knees bent and relaxed, and palpate (feel) the musculature of your posterior thigh. How do the muscles feel (taut or soft)? Now, stand up and bend over to touch your toes. Palpate the muscles of your posterior thigh again. How do they feel? The reason they feel differently is that, as you bend over to touch your toes, you stretch the hamstring muscles. This triggers a stretch reflex, which results in shortening (or tightening) of those muscles.

Technically, this reflex can be carried out without the help of the cerebral cortex and can be mediated solely by the spinal cord. We will see shortly, however, that the cortex is involved in even the simplest example of a stretch reflex—the patellar tendon (knee-jerk) reflex.

Procedure

1. With your partner seated, palpate the patellar tendon between the tibial tuberosity and the patella.

2. Tap this area with the flat end of a reflex hammer (sometimes a few taps are necessary to hit the right spot). What happens?

3. Now give your partner a difficult math problem to work (long division with decimals usually does the trick). As your partner works the problem, tap the tendon again. Is this response different from the original response? If yes, how, and why?

✔ Check Your Understanding

1. What might cause an absent or weak patellar tendon reflex?

2. Sometimes the reflex response is exaggerated—a phenomenon termed *hyperreflexia*. Do you think hyperreflexia would be caused by disorders of the central nervous system or of the peripheral nervous system? Explain your reasoning.

4 EXERCISE 4:
The Autonomic Nervous System_____

In the Pre-Lab Exercises, you learned about the *autonomic nervous system*—the largely involuntary branch of the peripheral nervous system charged with maintaining homeostasis in the face of changing conditions. The ANS, with its two branches, the *sympathetic* and the *parasympathetic* nervous systems, is truly one of the most important systems in the body, as it affects all of the other systems, from the cardiovascular and urinary to the integumentary systems.

The first branch of the ANS, the sympathetic nervous system (SNS), is often described in terms of the "fight or flight" response, in which the body prepares to respond to an emergency situation (e.g., running from a hungry alligator). This gives rise to a useful mnemonic: Your body is *sympathetic* to your emergency situation, so your *sympathetic* nervous system comes to the rescue.

Although this is a primary function of the SNS, it isn't the whole story. The SNS is activated by *any* excitement, emotion, or exercise, even something as simple as standing up. Similarly, the parasympathetic nervous system is discussed in terms of its "rest and recovery" role, in which it promotes functions associated with digestion, defecation, and diuresis (urine production). Note, however, that it begins taking over the moment the SNS response subsides. Each system works constantly to provide just the proper balance for any given situation. The following simple exercise will allow you to see the ANS in action.

Procedure

1. With your partner seated, take his/her resting pulse by placing two fingers on his/her radial pulse (on the lateral side of the wrist). Count the number of beats for 1 minute. Record the pulse in the chart below.

2. Obtain a stethoscope and listen to your partner's heart. How does the heartbeat sound (quiet, loud, etc.)?

3. Take your partner's resting blood pressure using a sphygmomanometer (blood pressure cuff) and a stethoscope (instructions for taking blood pressures are found in Unit 18) or an automated sphygmomanometer, if available. Record this information in the chart below.

4. Have your partner perform vigorous exercise for several minutes (running up and down stairs works well). Note that even minor exercise will produce blood pressure and pulse changes, but the changes are more visible with vigorous exercise (especially for those that are novices at taking blood pressures!).

5. Immediately after the vigorous exercise, repeat the pulse and blood pressure measurements, then listen to your partner's heart with a stethoscope. How does the heartbeat sound compared to the sounds that you heard at rest?

6. Have your partner rest seated for 5 minutes. Repeat the pulse and blood pressure measurements, and again listen to the heartbeat. Does the heart sound differently after rest?

ACTION	BLOOD PRESSURE	PULSE
Seated (at rest)		
Immediately after exercise		
After resting for 5 minutes		

7. Based upon the above data:

  a. In which situation(s) did the SNS dominate? How do you know?

  b. In which situation(s) did the PSNS dominate? How do you know?

✔ Check Your Understanding

1. Drugs used to treat hypertension (high blood pressure) often work by blocking different components of the SNS. Explain how inhibiting the SNS may act to lower blood pressure.

2. The disease *asthma* is characterized by increased airway resistance secondary to constriction of the bronchioles, airway inflammation, and excessive secretion of mucus. One of the main classes of drugs used to treat asthma, known as beta-two agonists, activates the beta-two receptors of the SNS on the bronchioles of the lungs. Explain how activating the SNS would treat an asthma attack.

15. Special Senses

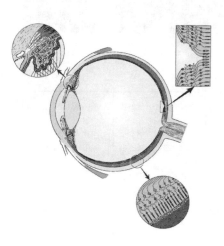

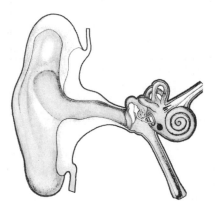

■ OBJECTIVES

Once you have completed this unit, you should be able to:

1. Identify structures of the eye on anatomical models and preserved or fresh specimens.
2. Describe the extraocular muscles that move the eyeball.
3. Compare the functions of the rods and cones.
4. Identify anatomical structures of the ear on anatomical models.
5. Perform tests of hearing and equilibrium.
6. Identify structures of the olfactory and taste senses on anatomical models.
7. Determine the relative concentration of cutaneous sensory receptors in different regions of the body.

■ MATERIALS

- Anatomical models: ear, eye, brain, tongue, and half-head
- Colored pencils
- Water-soluble marking pens
- Rulers
- Wooden applicator sticks
- Preserved or fresh cow eyeballs
- Dissecting equipment
- Tuning forks: 512 Hz
- Snellen chart
- Dark green and dark blue sheets of construction paper

PRE-LAB EXERCISES

Prior to coming to lab, complete the following exercises, using Chapter 11 in your lab atlas and your textbook for reference.

1 PRE-LAB EXERCISE 1: Anatomy of the Eye _____

Label and color-code Figure 15.1, illustrating the eye, using the terms from Exercise 1. Please note that all structures may not be visible on the diagrams. See Figures 11.1 and 11.3–11.5 in your lab atlas for reference.

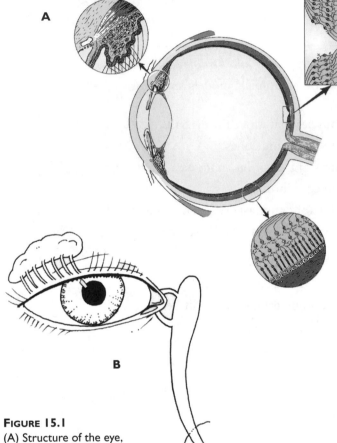

FIGURE 15.1
(A) Structure of the eye,
(B) external anatomy
and accessory structures
of the eye.

2 PRE-LAB EXERCISE 2: Extraocular Muscles_____

The six muscles that move the eyeball are called *extraocular muscles* (these were discussed briefly in Unit 14, with the cranial nerves). Fill in the following chart with the location, action, and cranial nerve innervation for each of the extraocular muscles.

MUSCLE	LOCATION	ACTION	CRANIAL NERVE INNERVATION
Superior rectus muscle			
Inferior rectus muscle			
Medial rectus muscle			
Lateral rectus muscle			
Superior oblique muscle			
Inferior oblique muscle			

3 PRE-LAB EXERCISE 3: Anatomy of the Ear _____

Label and color-code the structures of the ear in Figure 15.2, using the terms from Exercise 2. Please note that all structures may not be visible in the figure. See Figures 11.2, 11.7, and 11.9 in your lab atlas for reference.

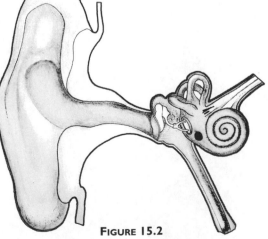

FIGURE 15.2
Anatomy of the ear.

Exercises

Sensation is broadly defined as awareness of changes in the internal and external environments. Sensation may be conscious or subconscious, depending on the destination of the sensory information. For example, there are receptors in certain blood vessels that detect blood pressure. This information is taken to the brain stem, which makes changes as necessary to ensure that blood pressure remains relatively constant. This information never makes it to the cerebral cortex, so you are not consciously aware of it. However, information that is eventually taken to the cerebral cortex (e.g., the taste of your food or the light level in a room) is something of which you are consciously aware. This type of sensation is called *perception*, and it is the focus of this unit.

The following exercises will ask you to examine the anatomy and physiology of the organs of perception, commonly called the *special senses*. The special senses include vision, hearing and equilibrium, taste, smell, and touch.

▮ EXERCISE 1: Anatomy of the Eye _____

The eye is a complex organ that consists of external structures such as the *eyelids* and *eyelashes*, accessory structures such as the *lacrimal gland*, and the *eyeball*. Many of the external and accessory structures of the eye protect the delicate eyeball.

Anteriorly, the eye is covered by the eyelids, or *palpebrae*. The eyelids meet medially and laterally at the medial and lateral *canthus*, respectively. The interior of the eyelids, as well as the much of the anterior eyeball, is covered with a thin mucous membrane called the *conjunctiva*. Within the eyelids are the *meibomian glands*, whose oily secretions help to keep the eye lubricated.

The two most prominent accessory structures of the eye are (1) the *lacrimal apparatus*, which produces and drains tears, and (2) the *extraocular muscles*. Refer to the Pre-Lab Exercises to review the locations, actions, and innervations of the extraocular muscles.

The eyeball is a hollow organ with three distinct tissue layers, or tunics:

1. *Fibrous tunic*: This outermost layer of the eyeball consists mostly of dense irregular connective tissue. It is avascular (lacks a blood supply), and consists of two parts:
   a. *Sclera*: Making up the posterior 5/6 of the fibrous tunic, this is the white part of the eyeball. It is white because of numerous collagen fibers that contribute to its thickness and toughness (in the same way that a joint capsule or a ligament is tough and white).
   b. *Cornea*: Making up the anterior 1/6 of the fibrous tunic, the cornea is clear, and it is one of the refractory media of the eyeball (it bends light coming into the eye).

2. *Vascular tunic*: Also called the *uvea*, this tunic carries most of the blood supply to the tissues of the eye. It is composed of three main parts:
   a. *Choroid*: The posterior part of the vascular tunic is made of the highly vascular choroid. The choroid is brown in color to prevent light scattering in the eye.
   b. *Ciliary body*: This structure, at the anterior aspect of the eye, is made chiefly of the *ciliary muscle*, which controls the shape of the lens.
   c. *Iris*: The most anterior portion of the uvea, the iris is the pigmented part of the eye. It consists of muscle fibers arranged around an opening called the *pupil*. As the fibers contract, the pupil changes size, either constricting or dilating.

3. *Sensory tunic*: This layer consists of the *retina* and the *optic nerve*. The retina is a thin, delicate structure that contains *photoreceptors* called *rods* and *cones*.
   a. The *rods* are scattered throughout the retina and are responsible for vision in dim light and for peripheral vision.
   b. *Cones* are concentrated at the posterior portion of the retina and are found in highest numbers in an area called the *macula lutea*. Cones are responsible for color and high-acuity vision.

Note that there are no rods or cones at the posteriormost aspect of the eyeball where the optic nerve leaves the eyeball. This location is called the *optic disc* or blind spot.

Another component of the eyeball is the *lens*, which allows for precise focusing of light on the retina. The lens divides the eyeball into the *anterior* and *posterior cavities* (sometimes called the anterior and posterior segments). The anterior cavity is filled with a watery fluid called *aqueous humor*, and the posterior contains a thicker fluid called *vitreous humor*. Both help to refract light onto the retina.

Identify the following structures of the eye and the eyeball on models and diagrams, using Pre-Lab Exercise 1 and Figures 11.1 and 11.3–11.5 in your lab atlas as a guide.

Structures of the Eyeball

1. Fibrous tunic
   a. Sclera
   b. Cornea

2. Vascular tunic (uvea)
   a. Choroid
   b. Ciliary body
   c. Suspensory ligaments
   d. Iris
   e. Pupil
3. Lens
4. Sensory tunic
   a. Retina
   b. Optic disc
   c. Macula lutea

   d. Fovea centralis
5. Optic nerve
6. Optic chiasma
7. Anterior cavity
   a. Anterior chamber
   b. Posterior chamber
   c. Aqueous humor
   d. Scleral venous sinus
8. Posterior cavity
   a. Vitreous humor
   b. Hyaloid canal

Accessory Structures

1. Palpebrae
2. Palpebral fissures
3. Medial and lateral canthi
4. Lacrimal apparatus
   a. Lacrimal gland
   b. Lacrimal caruncle
   c. Lacrimal canal
   d. Nasolacrimal duct
5. Conjunctiva

6. Extraocular muscles
   a. Superior oblique
   b. Inferior oblique
   c. Superior rectus
   d. Inferior rectus
   e. Medial rectus
   f. Lateral rectus

Model Inventory

As you examine the anatomical models and diagrams in lab, list them on the inventory below, and state which structures you are able to locate on each model.

MODEL/DIAGRAM	STRUCTURES IDENTIFIED

Eyeball Dissection

In this exercise you will examine the structures of the eyeball on a fresh or preserved eyeball. I promise that eyeball dissection isn't as gross as it sounds! As you dissect, use Figures 11.12–11.16 in your lab atlas as a guide.

Procedure

➤ NOTE: GOGGLES AND GLOVES ARE REQUIRED.

1. Examine the external anatomy of the eyeball and, below, list structures that you can identify.

2. Use scissors to remove the adipose tissue surrounding the eyeball. Identify the optic nerve.

3. Holding the eyeball at its anterior and posterior poles, use a sharp scalpel or scissors to make an incision in the frontal plane. Watch out, as aqueous and vitreous humor are likely to spill everywhere!

4. Complete the incision, separating the anterior and posterior portions of the eyeball. Take care to preserve the fragile retina—the thin, delicate yellow-tinted inner layer.

5. List the structures that you can identify in the anterior portion of the eyeball:

6. List the structures that you can identify in the posterior portion of the eyeball:

✔ Check Your Understanding

1. Corneal transplants may be performed with very little, if any, risk of transplant rejection by the immune system. Explain why this is so, taking into account the type of tissue found in the fibrous tunic of the eyeball. (*Hint:* Think about the location of the blood supply of the eye.)

2. The disease *macular degeneration* is characterized by a gradual loss of vision as a result of degeneration of the macula lutea. Considering the type of cells located in the macula lutea, which type of vision do you think a sufferer of macular degeneration would lose? Why?

3. What effect does the sympathetic nervous system have on the iris? What does it do to the size of the pupil? Considering the functions of the sympathetic nervous system, explain this effect.

Comparing the Distribution of Rods and Cones

Earlier, we discussed the unequal distribution of the photoreceptors in the retina: the rods and cones. In this exercise you will see (no pun intended) firsthand the differences in vision produced by the rods and the vision produced by the cones.

Procedure

1. On a small sheet of paper, write the phrase "Anatomy is fun" in your regular-size print.

2. Hold this piece of paper about 10 inches directly in front of your lab partner's eyes, and have your lab partner read the phrase.

   Can your partner read the phrase clearly?

   Which photoreceptors are producing the image?

3. Now hold the paper about 10 inches from your partner's peripheral vision field. Have him/her continue to stare forward and read the phrase.

   Can your partner read the phrase clearly?

   Which photoreceptors are producing the image?

4. For the next test, dim the lights in the room. Have your partner stand 20 feet in front of a Snellen eye chart and read the chart. Record the number of the smallest line that he/she can read (e.g., 20/40).

   Visual acuity:

5. With the lights still dimmed and your partner standing in the same place, hold a piece of dark green or dark blue construction paper over the Snellen chart. Ask your partner to identify the color of the paper you are holding:

   Paper color:

6. Repeat the above processes with the lights illuminated:

   Visual acuity:

   Paper color:

7. In which scenario was visual acuity and color vision better? Explain your findings.

Extraocular Muscles

Recall that the eyeball is moved by a set of six extraocular muscles. In this exercise, you will determine which extraocular muscles are responsible for moving the eyeballs in each direction. Refer to Pre-Lab Exercise 2 for assistance.

1. With your lab partner seated, trace a horizontal line in the air about one foot in front of your partner's eyes, moving from right to left. Have your partner follow your finger without moving his/her head. Which extraocular muscles produce the movements you see for each eye?

   *Right eye*_____

   *Left eye*_____

2. Now trace a diagonal line, starting at the upper right corner and moving to the lower left corner. Have your partner follow your finger again. Which extraocular muscles produce the movements that you see for each eye?

   *Right eye*_____

   *Left eye*_____

3. Again have your partner follow your finger, but trace a horizontal line from left to right. Which extraocular muscles produce the movements that you see for each eye?

   *Right eye*_____

   *Left eye*_____

4. Finally, trace another diagonal, this time from the lower left to the upper right, and have your partner follow along (you should have traced out an hourglass shape overall). Which extraocular muscles produce the movements that you see for each eye?

   *Right eye*_____

   *Left eye*_____

2 EXERCISE 2: Anatomy of the Ear _____

The ear contains structures for both hearing and equilibrium. It is divided into three regions: the outer, middle, and inner ear.

1. *Outer ear*: The outer ear begins with the *auricle*, or pinna, a shell-shaped structure composed of elastic cartilage that surrounds the opening to the *external auditory canal*. The external auditory canal extends about 2.5 cm into the temporal bone, where it ends in the *tympanic membrane*, which divides the outer ear from the middle ear.

2. *Middle ear*: The middle ear is a small cavity within the temporal bone that houses tiny bones called the auditory *ossicles*—the *malleus*, *incus*, and *stapes*. The ossicles transmit vibrations from the tympanic membrane through the middle ear, and to the inner ear through a structure called the *oval window*. An additional structure in the middle ear is the *pharyngotympanic tube* (also called the Eustachian or auditory tube). It connects the middle ear to the pharynx (throat) and equalizes pressure in the middle ear.

3. *Inner ear*: The inner ear contains the sense organs for hearing and equilibrium. It consists of bony cavities, called the *bony labyrinth*, which are filled with a fluid called *perilymph*. Within the perilymph is a series of membranes called the *membranous labyrinth*, which contains a thicker fluid called *endolymph*.

   a. *Vestibule*: The vestibule is an egg-shaped bony cavity that houses two structures responsible for equilibrium: the *saccule* and the *utricle*. Both structures transmit impulses down the vestibular portion of the vestibulocochlear nerve.

   b. *Semicircular canals*: Situated at right angles to one another, the semicircular canals house the *semicircular ducts* and the *ampulla*, which work together with the vestibule to maintain equilibrium. Their orientation allows them to sense rotational movements of the head and body. Like the saccule and utricle, the semicircular ducts and the ampulla also transmit impulses down the vestibular portion of the vestibulocochlear nerve.

   c. *Cochlea*: This spiral bony canal contains the *organ of Corti*, whose specialized *hair cells* transmit sound impulses to the cochlear portion of the vestibulocochlear nerve.

Identify the following parts of the ear on models and diagrams, using Pre-Lab Exercise 3 and Figures 11.2, 11.7, and 11.9 in your lab atlas as a guide.

Structures of the Ear

1. Outer ear
   a. Auricle (pinna)
   b. External auditory canal

2. Middle ear
   a. Tympanic membrane
   b. Ossicles
      (1) Malleus
      (2) Incus
      (3) Stapes
   c. Oval window
   d. Round window

   e. Pharyngotympanic (auditory or Eustachian) tube

3. Inner ear
   a. Vestibule
      (1) Saccule
      (2) Utricle
   b. Semicircular canals
      (1) Semicircular duct
      (2) Ampulla
   c. Cochlea
      (1) Organ of Corti

(2) Cochlear duct   (1) Vestibular nerve

d. Vestibulocochlear nerve   (2) Cochlear nerve

Model Inventory

As you view the anatomical models and diagrams in lab, list them on the inventory below, and state which structures you are able to locate on each model.

MODEL/DIAGRAM	STRUCTURES IDENTIFIED

Weber and Rinne Tests

When evaluating a subject's ability to hear, a distinction is made between two possible types of hearing loss.

1. *Conductive hearing loss*: results from interference of sound conduction through the outer and/or middle ear.

2. *Sensorineural hearing loss*: results from damage to the inner ear or the vestibulocochlear nerve.

The Weber and Rinne tests use tuning forks to distinguish between these types of hearing loss. Tuning forks, which vibrate at specific frequencies when struck, are placed directly on the bones of the skull to evaluate bone conduction—the ability to hear the vibrations transmitted through the bone. The forks are then held near the ear, not touching bone, to evaluate air conduction—the ability to hear the vibrations transmitted through the air.

Procedure: Weber Test

1. Obtain a tuning fork with a frequency of 500–1000 Hz (cycles per second).

2. Holding the tuning fork by the base, strike it lightly with a mallet or tap it on the edge of the table. The fork

should begin ringing softly. If it is ringing too loudly, grasp the tines to stop it from ringing and try again.

3. Place the base of the vibrating tuning fork on the midline of your partner's head.

4. Ask your partner if the sound is heard better in one ear or if the sound is heard equally in both ears. If the sound is heard better in one ear, this is called lateralization.

Was the sound lateralized? If yes, to which ear?

5. To illustrate what it would sound like if the sound were lateralized, have your partner place his/her finger in one ear. The ear with the finger blocking the external auditory canal represents an ear with conduction deafness. Repeat the test.

a. In which ear was the sound heard better?

b. If a patient were to have conduction deafness, in which ear do you think that the sound would be heard the most clearly (the deaf ear or the good ear)? _____ Why? (If you are confused, think about your results when one ear was plugged.)

c. If the patient were to have sensorineural deafness, in which ear do you think that the sound would be best heard?

Procedure: Rinne Test

1. Strike the tuning fork lightly to start it ringing.

2. Place the base of the tuning fork on your partner's mastoid process.

3. Begin timing the interval during which your partner can hear the sound. Your partner will have to tell you when he/she can no longer hear the ringing.

Time interval in seconds:

4. Once your partner can no longer hear the ringing, quickly move the still-vibrating tuning fork 1–2 cm from the external auditory canal.

5. Time the interval from the point when you moved the tuning fork in front of the external auditory canal to when your partner can no longer hear the sound.

Time interval in seconds:

Which situation tested bone conduction?

Which situation tested air conduction?

6. Typically, the air-conducted sound is heard twice as long as the bone-conducted sound. For example, if the bone-conducted sound was heard for 15 seconds, the air-conducted sound should be heard for 30 seconds.

Were your results normal?

What would be indicated if the bone-conducted sound were heard for longer than the air-conducted sound? *(Hint: Think about your results from the Weber test.)*

Romberg Test

A common and simple test of equilibrium is the Romberg test, in which the person being tested is asked to stand still first with the eyes open and then with the eyes closed. Under normal conditions, the vestibular apparatus should be able to maintain equilibrium in the absence of visual input. If the vestibular apparatus is impaired, however, the brain relies on visual cues to maintain balance.

Procedure

1. Have your partner stand erect with the feet together and the arms at the sides in front of a chalkboard.
2. Use chalk to draw lines on the board on either side of your partner's torso. These lines are for your reference in the next part.
3. Have your partner stand in front of the chalkboard for 1 minute, staring forward with his/her eyes open. Use the lines on either side of his/her torso to note how much he/she sways as she stands. Below, record the amount of side-to-side swaying (i.e., minimal or significant):

4. Now have your partner stand in the same position with his/her eyes closed for 1 minute. Again note the amount of side-to-side swaying, using the chalklines for reference.

Was the amount of swaying more or less with his/her eyes closed?

Why do you think this is so?

What do you predict would be the result for a person with an impaired vestibular apparatus? Explain.

✔ Check Your Understanding

1. What signs and symptoms would you expect to see from otitis interna (inner ear infection)? Why?

2. Explain why infectious otitis media (inflammation of the middle ear) may result in a concomitant pharyngitis (inflammation of the throat).

3. Otosclerosis is a condition that results in irregular ossification (bone formation) around the stapes. Would you expect this to result in conductive or sensorineural hearing loss? What results would you expect from the Rinne and Weber tests?

3 EXERCISE 3: Olfactory and Taste Senses _____

Both olfaction and taste are sometimes referred to as the *chemosenses*, because they both rely on chemoreceptors to relay information about the environment to the brain.

The chemoreceptors of the olfactory sense are located in a small patch in the roof of the nasal cavity called the *olfactory epithelium*. The olfactory epithelium contains bipolar neurons called olfactory receptor cells. Their axons penetrate the holes in the cribriform plate to synapse on the olfactory bulb, which then sends the impulses down the olfactory nerves (cranial nerve I) to the olfactory cortex.

Taste receptors are located on taste buds that are housed on projections from the tongue called *papillae*. There are three types of papillae: filiform, fungiform, and circumvallate. Of the three, only fungiform and circumvallate papillae house taste buds. Fungiform papillae are scattered over the surface of the tongue, whereas the large circumvallate papillae are located at the posterior aspect of the tongue, arranged in a V-shape.

Identify the following structures of the olfactory and taste senses on anatomical models and charts.

Structures of Olfaction

1. Nasal cavity
   a. Nasal conchae
   b. Nasal septum
2. Olfactory epithelium
   a. Olfactory receptor cells

3. Olfactory bulbs
4. Olfactory nerves
5. Cribriform plate

Structures of Taste

1. Papillae
   a. Fungiform papillae
   b. Circumvallate papillae
2. Taste buds

As you view the anatomical models and diagrams in lab, list them on the inventory below and state which structures you are able to locate on each model.

Model/Diagram	Structures Identified

✔ Check Your Understanding

1. The chemoreceptors of the tongue respond to four basic taste sensations: sweet, sour, bitter, and salty. The sweet sensation is created by the presence of monosaccharides (single-sugar residues), which is why a cookie tastes sweet. Why do you think a potato doesn't taste sweet?

2. The remainder of the chemoreceptors respond to metal ions (salty tastes), alkaloids (bitter tastes), and acids (sour tastes). Which type of receptor do you think is most sensitive? Why? (*Hint: One of the three chemicals is present in many plant poisons.*)

4 EXERCISE 4: Cutaneous Sensation _____

Sensory receptors in the skin respond to different stimuli, including temperature, touch, and pain. These receptors are not distributed throughout the skin equally but instead are concentrated in certain regions of the body. The following experiments will allow you to determine the relative distribution of the receptors for touch in the skin by performing two tests: the error of localization and two-point discrimination.

Error of Localization

Every region of the skin corresponds to a specific part of the somatosensory association area in the cerebral cortex. Some regions are better represented than others and therefore are capable of localizing stimuli with greater precision than less well-represented areas. The error of localization (also called tactile localization) tests the ability to determine the precise location of the skin that has been touched, and therefore demonstrates how well-represented each region of the skin is in the cerebral cortex.

Procedure

1. Have your partner sit with his/her eyes closed.

2. Use a water-soluble marking pen to place a mark on your partner's anterior forearm.

3. Using a different color of marker, have your partner, still with his/her eyes closed, place a mark as close as possible to where he/she believes the original spot is located.

4. Use a ruler to measure the distance between the two points in millimeters. This is your error of localization.

5. Repeat this procedure for each of the following locations:
   a. Anterior thigh
   b. Face
   c. Palm of hand
   d. Fingertip

6. Record your data in the chart below.

REGION	ERROR OF LOCALIZATION (MM)
Anterior forearm	
Anterior thigh	
Face	
Palm of hand	
Fingertip	

Two-Point Discrimination

This test assesses the ability to perceive the number of stimuli ("points") that are placed on the skin. Areas that have a higher density of touch receptors have a better ability to distinguish between multiple stimuli than those with fewer touch receptors.

Procedure

1. Have your partner close his/her eyes.

2. Place the ends of two wooden applicator sticks close together (they should be nearly touching) on your partner's skin on the anterior forearm. Ask your partner how many points he/she can discriminate: one or two.

3. If he/she can sense only one point, move the sticks farther apart. Repeat this procedure until your partner can distinguish two separate points touching his/her skin.

4. Use a ruler to measure the distance between the two sticks in millimeters. This is your 2-point discrimination.

5. Repeat this procedure for each of the following locations:
   a. Anterior thigh
   b. Face (around the lips and/or eyes)
   c. Vertebral region
   d. Fingertip

6. Record your data in the chart below:

REGION	TWO-POINT DISCRIMINATION (MM)
Anterior forearm	
Anterior thigh	
Face	
Vertebral region	
Fingertip	

✔ Check Your Understanding

1. What results did you expect for each test? Explain.

2. Did your observations agree with your expectations? Interpret your results.

3. A map called the *sensory homunculus* is used to diagram the amount of the cerebral cortex that is dedicated to one specific area of the body. On the sensory homunculus, areas of the body that correspond to greater areas of the cortex are represented as proportionately larger than those areas that correspond to smaller areas of the cortex. Which areas of the body do you think would be larger on the homunculus? Which areas do you think would be smaller?

16. Anatomy of the Heart

■ **OBJECTIVES**

Once you have completed this unit, you should be able to:

1. Identify anatomical structures of the heart.
2. Trace the pathway of blood flow through the heart.
3. Describe and observe the histology of cardiac tissue.
4. Perform a dissection of the heart on preserved or fresh hearts.

■ **MATERIALS**

- Anatomical models: hearts, torsos
- Microscope slides: cardiac tissue
- Laminated outline of the heart
- Water-soluble marking pens
- Fresh or preserved hearts
- Dissecting trays, kits
- Colored pencils

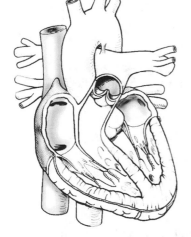

PRE-LAB EXERCISES

Prior to coming to lab, use your text and lab atlas to complete the following exercises, referencing Chapter 12 in your lab atlas.

1 PRE-LAB EXERCISE 1: Anatomy of the Heart _____

Figure 16.1 presents diagrams of the heart, from the anterior view and a frontal section. Use colored pencils to color-code the diagrams, and label them with the structures indicated in Exercise 1.

> ➤ **NOTE:** IT IS USEFUL TO USE RED TO COLOR-CODE THE VESSELS CARRYING OXYGENATED BLOOD, AND BLUE FOR THOSE CARRYING DEOXYGENATED BLOOD.

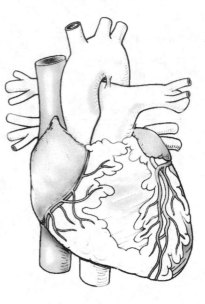

A

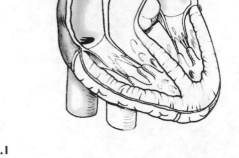

B

FIGURE 16.1
Diagrams of the heart: (A) anterior, and (B) frontally sectioned views.

2 PRE-LAB EXERCISE 2: Pathway of Blood Flow Through the Heart __

The following questions pertain to the pathway of blood flow through the heart.

1. Regarding veins:
   a. Where do veins carry blood?

   b. Is this blood generally oxygenated or deoxygenated?

   c. Are there any exceptions to this rule? If yes, where?

2. Regarding arteries:
   a. Where do arteries carry blood?

   b. Is this blood generally oxygenated or deoxygenated?

   c. Are there any exceptions to this rule? If yes, where?

3. Where does each atrium pump blood when it contracts?
   a. Right atrium:

   b. Left atrium:

4. Where does each ventricle pump blood when it contracts?
   a. Right ventricle:

   b. Left ventricle:

EXERCISES

The primary function the cardiovascular system is the transport of substances, including oxygen, carbon dioxide, nutrients, and white blood cells, through the body using the blood as a transport vehicle. The pump that drives the movement of the blood around the body is the heart. The heart is a remarkable organ, tirelessly beating over 100,000 times per day to move over 8,000 liters of blood.

In this unit, we will examine the structures of this remarkable organ, including the pathway of blood flow through the heart and the histology of cardiac muscle. In the third exercise you will dissect and examine a fresh or preserved heart.

EXERCISE 1: Structures of the Heart

The heart consists of four hollow chambers that receive blood from and eject blood into the major vessels of the body. The chambers are bordered by valves that help to prevent blood from flowing backward though the heart. Refer to your Pre-Lab Exercises to review the anatomy of the chambers, valves, and major vessels.

The heart lies in the *mediastinum* and is, on average, the size of a fist. When the cavity is opened, however, the heart is not immediately visible because it is encased in a secondary cavity called the *pericardial cavity*. The pericardial cavity has two serous membranes, between which is found a layer of serous fluid:

1. *Fibrous pericardium* (see Figure 12.7 in your lab atlas): the fibrous outer layer that anchors the heart to the surrounding structures of the mediastinum. It lacks elastic fibers and therefore is not very distensible, which prevents the heart from overfilling.

2. *Serous pericardium*: the thin, delicate inner serous membrane. It consists of two layers:
   a. *Parietal pericardium*: the layer that is functionally fused to the fibrous pericardium.
   b. *Visceral pericardium*: the layer, also called the *epicardium*, that is functionally fused to the heart muscle (the myocardium). The pericardial cavity and serous fluid are found between the parietal and the visceral pericardium.

Because the heart is an organ, it must have more than one type of tissue. Therefore, in addition to the layers of the pericardium, the heart has two other tissue layers:

1. *Myocardium*: the actual heart muscle, which is the thickest layer, consisting of cardiac muscle cells wrapped around a meshwork of collagen and elastic fibers called the *fibrous skeleton*.

2. *Endocardium*: a simple squamous epithelial tissue layer that lines the inner surface of the heart and the valves; it is continuous with the inner lining of blood vessels that enter and leave the heart, where it is called the *endothelium*.

Additional structures of the heart that will be examined in this lab period are the vessels of the *coronary circulation* (see Figure 12.3 in your lab atlas). A common misconception is that the heart requires no blood vessels of its own, as it contains blood within its chambers. The nutrients in blood, however, would not be able to diffuse far enough down to nourish the cells of the myocardium, and the chambers on the right side contain deoxygenated blood, which is depleted of nutrients and oxygen. Therefore, the heart necessitates its own set of blood vessels, which together are known as the *coronary vessels*.

Two *coronary arteries* (the right and left coronary arteries) branch off the base of the aorta, to serve the myocardium. The coronary arteries are drained by a set of *coronary veins*, which drain into a common vessel on the posterior aspect of the heart called the *coronary sinus*, which in turn drains into the base of the right atrium. When a coronary artery is blocked, the reduced blood flow to the myocardium can result in hypoxic injury and death to the tissue, termed a *myocardial infarction* (commonly called a heart attack).

Following is a list of structures that we will identify in lab. As you identify the structures on models and diagrams, use your Pre-Lab Exercises and Figures 12.3, 12.4, 12.6, 12.7, and 12.9–12.17 in your lab atlas for reference

Anatomical Structures of the Heart

1. General structures
   a. Mediastinum
      (1) Pericardial cavity
   b. Pericardium
      (1) Parietal pericardium
      (2) Visceral pericardium (epicardium)
   c. Myocardium
   d. Endocardium
   e. Apex of the heart
   f. Base of the heart

2. Structures of the atria
   a. Right atrium
   b. Left atrium
   c. Right auricle
   d. Left auricle

   e. Interatrial septum

   f. Fossa ovalis

   g. Pectinate muscles

   h. Opening of the coronary sinus

   i. Sinoatrial node

3. Structures of the ventricles

   a. Right ventricle

   b. Left ventricle

   c. Interventricular septum

   d. Anterior interventricular sulcus

   e. Posterior interventricular sulcus

   f. Trabeculae carneae

4. Atrioventricular valves

   a. Tricuspid valve

   b. Mitral valve

5. Semilunar valves

   a. Pulmonic valve

   b. Aortic valve

6. Valvular structures

   a. Chordae tendineae

   b. Papillary muscles

7. Great vessels

   a. Superior vena cava

   b. Inferior vena cava

   c. Pulmonary trunk (artery)

   d. Right and left pulmonary arteries

   e. Pulmonary veins

   f. Aorta

   g. Ligamentum arteriosum

8. Coronary arteries

   a. Right coronary artery

   b. Marginal artery

   c. Left coronary artery

   d. Anterior interventricular artery

   e. Circumflex artery

   f. Posterior interventricular artery

9. Coronary veins

   a. Great cardiac vein

   b. Middle cardiac vein

   c. Coronary sinus

Model Inventory

As you view the anatomical models and torsos in lab, list them on the inventory in the next column, and state which of the above structures you are able to locate on each model.

MODEL	STRUCTURES IDENTIFIED

Tracing the Pathway of Blood Flow Through the Heart

Use water-soluble markers and a laminated outline of the heart to trace the pathway of blood as it flows through the heart and pulmonary circulation. Use a blue marker to indicate areas that contain deoxygenated blood and a red marker to indicate areas that contain oxygenated blood. If no laminated outline is available, use Figure 16.2.

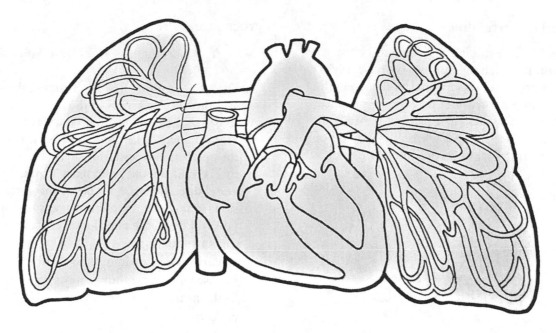

FIGURE 16.2
Diagram of frontal section of heart, with lungs and pulmonary circulation.

✔ Check Your Understanding

1. Rheumatic heart disease is a disorder of the immune system in which the valves of the heart are attacked. Predict the effects this valvular disease would have on the valves themselves, the myocardium, and other tissues (e.g., how would a defective aortic valve affect a person's stamina?).

2. When the pericardium fills with blood, it produces a condition called *cardiac tamponade*, which can be rapidly lethal. Why is this condition so dangerous, taking into account the structure of the fibrous pericardium?

2 EXERCISE 2: Cardiac Muscle Histology

Recall from Unit 5 that cardiac muscle tissue is striated like skeletal muscle tissue but otherwise is quite different. Following are some important differences:

- Cells of cardiac muscle are short and fat rather than long and thin like skeletal muscle.

- Cardiac muscle cells typically are uninucleate.

- These cells contain specialized adaptations called *intercalated discs*, which appear as dark lines that run parallel to the striations. They function to hold adjacent cardiac cells tightly together so that the heart muscle beats as a unit, and to allow the cells to communicate chemically and electrically.

Examine a prepared slide of cardiac muscle tissue on high power. Use colored pencils to draw what you see, and label the structures indicated. Use Figure 3.37 in your lab atlas for reference.

Cardiac Muscle Tissue

Label the following structures:

1. Intercalated discs

2. Striations

3. Nucleus

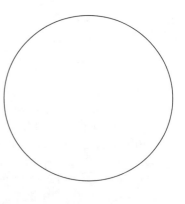

✔ Check Your Understanding

1. Skeletal muscle cells exhibit a phenomenon known as *recruitment*, in which the number of muscle cells recruited to contract is proportional to the strength of muscle contraction needed. In this way, we produce a small contraction to pick up a piece of paper and a larger contraction to pick up a textbook. Would you expect to see recruitment in cardiac muscle tissue? Why or why not?

2. Cardiac muscle cells are uninucleate, whereas skeletal muscle cells are multinucleate. Why do you think cardiac cells lack multiple nuclei?

3 EXERCISE 3: Heart Dissection

In this exercise we will examine a preserved heart or a fresh heart, which usually is from a sheep or cow. You will notice that certain structures, such as the pericardium and the chordae tendineae, are easier to visualize on a real specimen and other structures, such as the coronary vessels, are more difficult to see on a real specimen. Follow the procedure below to best find the structures indicated. You also may want to use Figure 12.35 in your lab atlas for reference.

Procedure

> ➤ NOTE: SAFETY GLASSES AND GLOVES ARE REQUIRED.

1. Orient yourself by first determining the superior aspect and the inferior aspect of the heart. The superior aspect (base) of the heart is the broad, flat end, and the inferior aspect (apex) is the pointy tip. Now orient yourself to the anterior and posterior sides. The easiest way to do this is to locate the pulmonary trunk. The trunk is the vessel directly in the middle of the anterior side. Find the side from which the pulmonary trunk originates, and you will be on the anterior side. Structures to locate at this time are:
   a. Parietal pericardium (may not be attached)
   b. Visceral pericardium (shiny layer over the surface of the heart)
   c. Aorta
   d. Pulmonary trunk
   e. Superior vena cava
   f. Inferior vena cava
   g. Pulmonary veins
   h. Ventricles
   i. Atria

   Finding the coronary vessels tends to be difficult because the superficial surface of the heart is covered with adipose tissue. To see the coronary vessels, careful dissection of the adipose tissue is necessary.

2. Locate the superior vena cava. Insert scissors or a scalpel into the superior vena cava, and cut down into the right atrium. Before moving onto Step 3, note the structure of the tricuspid valve and draw it below.

3. Once the right atrium is exposed, continue the cut down into the right ventricle. Structures to locate at this time include:
   a. Tricuspid valve
   b. Chordae tendineae
   c. Papillary muscles
   d. Myocardium
   e. Endocardium (shiny layer on the inside of the heart)

4. Insert the scissors into the pulmonary trunk. Note the structure of the pulmonic valve and draw it below:

5. Insert the scissors into a pulmonary vein. Cut down into the left atrium. Note the structure of the mitral valve, and draw it below:

6. Continue the cut into the left ventricle. Note at this time the thickness of the left ventricle. Compare it to the thickness of the right ventricle:

7. Insert the scissors into the aorta. Extend the cut until you can visualize the aortic valve. Draw the aortic valve below:

✔ Check Your Understanding

1. Explain the differences in size of the myocardium of the right ventricle and the left ventricle.

2. Note the smooth texture of the valves and how the leaflets fit together. How do you think the function of the valves would be affected if the valves were tough or filled with calcium deposits? Explain.

17. Blood Vessel Anatomy

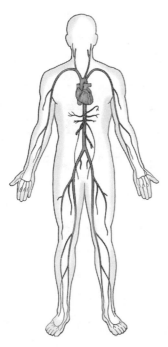

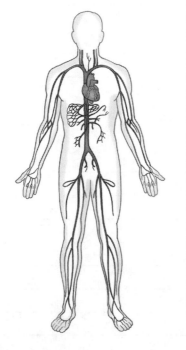

■ OBJECTIVES

Once you have completed this unit, you should be able to:

1. Identify arteries and veins on anatomical models.

2. Describe the organs and regions supplied and drained by each artery and vein.

3. Describe the unique blood flow patterns through the brain and the hepatic portal system.

4. Trace the pathway of blood flow through various arterial and venous circuits.

5. Describe the histological differences between arteries and veins on microscope slides.

■ MATERIALS

- Anatomical models: torsos, blood vessels
- Microscope slides: artery, vein, elastic artery
- Colored pencils
- Laminated outline of the body
- Water-soluble marking pens

PRE-LAB EXERCISES

Prior to coming to lab, complete the following exercises, using Chapter 12 in your lab atlas and your textbook for reference.

1 PRE-LAB EXERCISE 1: Arteries

Describe the location of the arteries in the following chart, and the organ and/or region that each artery supplies.

ARTERY	LOCATION	ORGAN/REGION SUPPLIED
Brachiocephalic artery		
Common carotid artery		
External carotid artery		
Temporal artery		
Internal carotid artery		
Vertebral artery		
Basilar artery		
Circle of Willis		
Anterior cerebral artery		
Middle cerebral artery		
Posterior cerebral artery		
Subclavian artery		
Axillary artery		
Brachial artery		
Radial artery		
Ulnar artery		
Celiac trunk		
Hepatic artery		
Left gastric artery		
Splenic artery		
Renal artery		
Superior mesenteric artery		
Inferior mesenteric artery		
Common iliac artery		
Internal iliac artery		
External iliac artery		
Femoral artery		
Popliteal artery		
Posterior tibial artery		
Anterior tibial artery		
Dorsalis pedis artery		

2 PRE-LAB EXERCISE 2:
Veins

Describe the location of the following veins, and the organ and/or region that each vein drains.

ARTERY	LOCATION	ORGAN/REGION SUPPLIED
Dural sinuses		
Internal jugular vein		
External jugular vein		
Brachiocephalic vein		
Ulnar vein		
Radial vein		
Median cubital vein		
Brachial vein		
Basilic vein		
Cephalic vein		
Subclavian vein		
Hepatic vein		
Hepatic portal vein		
Inferior mesenteric vein		
Splenic vein		
Superior mesenteric vein		
Renal vein		
Posterior tibial vein		
Anterior tibial vein		
Popliteal vein		
Femoral vein		
External iliac vein		
Internal iliac vein		
Common iliac vein		

3 PRE-LAB EXERCISE 3:
Arterial Anatomy

Label the diagrams in Figure 17.1 with the arteries from Exercise 1. Please note that all vessels may not be visible on the diagrams. Use colored pencils to trace the routes of blood flow. See Chapter 12 in your atlas for reference.

A

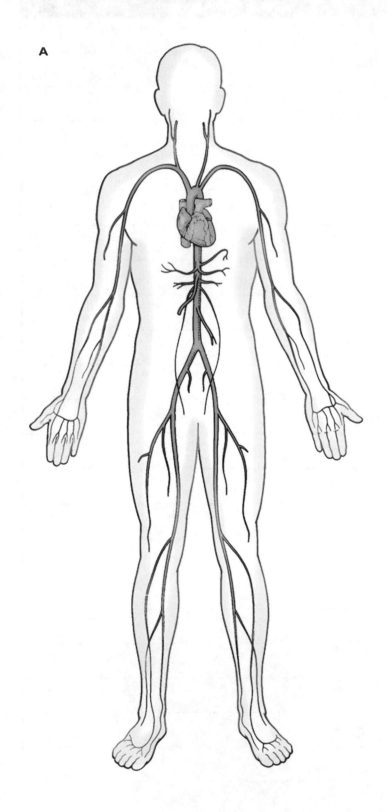

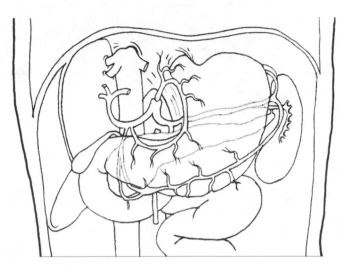

B

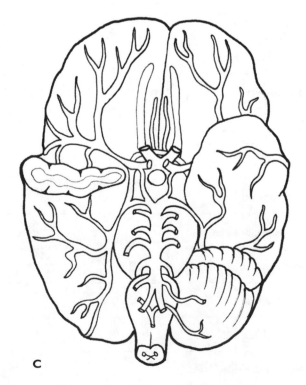

C

FIGURE 17.1

Diagrams of (A) arterial circuit, (B) abdominal arteries, (C) arterial supply to the brain.

4 PRE-LAB EXERCISE 4:
Venous Anatomy

Label the diagrams in Figure 17.2 with the veins from Exercise 1. Please note that all vessels may not be visible on the diagrams. Use colored pencils to trace the routes of blood flow. See Chapter 12 in your atlas for reference.

A

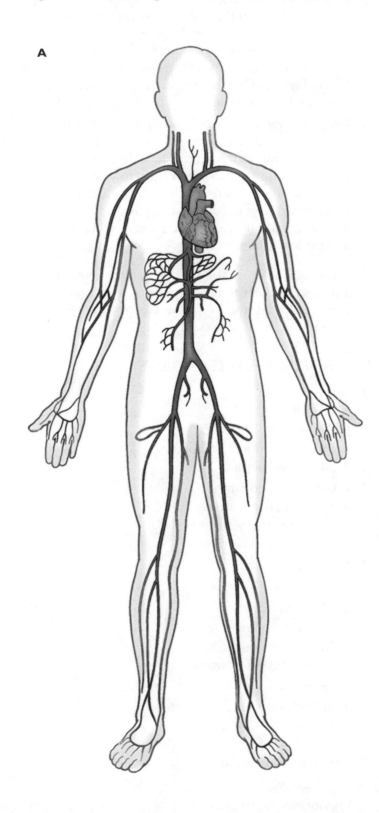

B

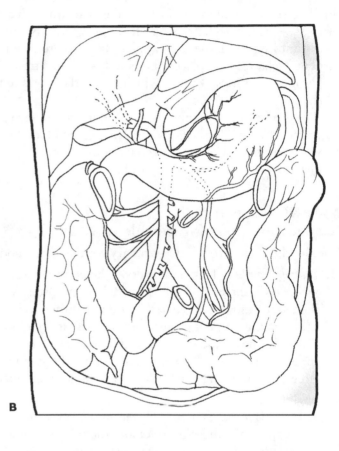

C

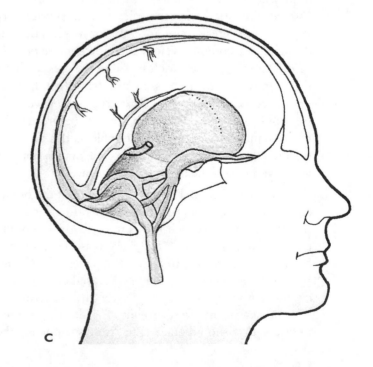

FIGURE 17.2
Diagrams of (A) venous circuit, (B) abdominal veins, (C) dural sinuses.

EXERCISES

The blood vessels are a closed system of tubes that carry blood around the body. With the heart as the pump, the blood vessels carry blood away from the heart through a series of *arteries,* which branch to form progressively smaller vessels until we reach the tiny capillary beds where gas and nutrient exchange take place. The blood is drained from the capillaries via a series of veins, which return the blood to the heart.

The exercises in this unit involve you in identifying the major blood vessels in the body and tracing the blood flow through the body. In the final exercise you will investigate the histology of the blood vessel wall.

EXERCISE 1:
Identification of Major Blood Vessels _____

The blood travels through two primary circuits in the body:

1. *Systemic circuit*: This circuit serves the body tissues, where oxygenated blood is delivered via the arteries to the tissues and deoxygenated blood is returned to the heart via the veins. It begins with the largest artery in the body, the *aorta*, and ends with two large veins, the superior and inferior *venae cavae*. Within the systemic circuit are two unique patterns of blood flow that merit mention:

   a. *Hepatic portal system*: The venous blood from the organs of the digestive tract and the spleen do not drain into the inferior vena cava, as would be expected. Instead, the blood from the splenic vein, gastric veins, superior mesenteric vein, and inferior mesenteric vein drains into a common vein, called the *hepatic portal vein*. Here the nutrient-rich blood percolates through the liver, where it is processed and detoxified. In this way, everything that we ingest (excepting, as we will see, lipids) must travel through the liver before entering the systemic circulation. Once the blood has filtered through the hepatic portal system, it exits via hepatic veins, and enters the inferior vena cava.

   b. *Cerebral circulation*: The arterial supply of the brain comes primarily from the internal carotid arteries and the basilar artery (which forms from the fusion of the two vertebral arteries). Once these vessels enter the brain, they contribute to a structure called the *Circle of Willis* (also called the cerebral arterial circle). This circle, composed of a set of anterior and posterior communicating arteries, provides for alternate routes of circulation should one of the arteries supplying the

brain become blocked. The venous system draining the capillaries of the brain is also unique. Venous blood from the brain does not simply drain into one vein and exit the head, but instead drains into a set of thin-walled veins called the *dural sinuses*. The inferior dural sinus, called the *transverse sinus*, drains into the internal jugular vein.

2. *Pulmonary circuit*: The pulmonary circuit transports deoxygenated, carbon dioxide-rich blood from the right side of the heart to the lungs, where the blood drops off carbon dioxide and picks up oxygen to be returned to the left side of the heart and the systemic circuit. It begins with the pulmonary trunk, which leaves the right ventricle, and ends with the pulmonary veins that drain into the left atrium.

The following is a list of blood vessels that you will cover in this lab. Refer to Pre-Lab Exercises 1-4, and Chapters 12, 19, 20, and 21 in your lab atlas to review the locations of each of these vessels.

Arteries of the Trunk, Head, and Neck

1. Aorta
   a. Ascending aorta
   b. Aortic arch
   c. Descending (thoracic) aorta
   d. Abdominal aorta
2. Common carotid artery
   a. External carotid artery
      (1) Temporal artery
   b. Internal carotid artery
3. Vertebral artery
4. Basilar artery
5. Circle of Willis
   a. Anterior cerebral artery
   b. Middle cerebral artery
   c. Posterior cerebral artery
6. Celiac trunk
   a. Splenic artery
   b. Left gastric artery
   c. Hepatic artery
7. Superior mesenteric artery
8. Renal artery
9. Inferior mesenteric artery

Arteries of the Limbs

1. Brachiocephalic artery
2. Subclavian artery
3. Axillary artery
4. Brachial artery
5. Radial artery
6. Ulnar artery
7. Common iliac artery
   a. Internal iliac artery

b. External iliac artery
8. Femoral artery
9. Popliteal artery
   a. Anterior tibial artery
      (1) Dorsalis pedis artery
   b. Posterior tibial artery

Veins of the Trunk, Head, and Neck

1. Superior vena cava
2. Inferior vena cava
3. Dural sinuses
   a. Superior sagittal sinus
   b. Cavernous sinus
   c. Transverse sinus
4. Internal jugular vein
5. External jugular vein
6. Brachiocephalic vein

7. Hepatic veins
8. Hepatic portal vein
9. Splenic vein
10. Superior mesenteric vein
11. Inferior mesenteric vein
12. Gastric vein
13. Renal vein

Veins of the Limbs

1. Ulnar vein
2. Radial vein
3. Median cubital vein
4. Brachial vein
5. Basilic vein
6. Cephalic vein
7. Axillary vein
8. Subclavian vein

9. Anterior tibial vein
10. Posterior tibial vein
11. Popliteal vein
12. Femoral vein
13. External iliac vein
14. Internal iliac vein
15. Common iliac vein

Pulmonary Circuit

1. Pulmonary trunk
2. Right and left pulmonary arteries
3. Pulmonary arteriole
4. Pulmonary venule
5. Pulmonary veins

Model Inventory

As you view the anatomical models, torsos, and charts in lab, list them on the inventory in the next column, and state which of the above structures you are able to locate on each model.

MODEL/DIAGRAM	STRUCTURES IDENTIFIED

✔ Check Your Understanding

1. Certain drugs are unable to be given by mouth because their hepatic metabolism is so great that the entire dose of the drug is destroyed in the liver before it ever reaches the general circulation. These same drugs can be given by injection, either intravenously or intramuscularly. Explain this, considering the blood flow through the hepatic portal system.

2. If a blood clot were to lodge in one of the anterior or posterior communicating arteries of the Circle of Willis, would this cause a significant defect? Why or why not?

3. Typically on an anatomical model, blood vessels carrying deoxygenated blood are colored blue, and those carrying oxygenated blood are colored red. On a model of the pulmonary vessels, which vessels would be blue? Which would be red? Why?

2 EXERCISE 2: Time to Trace!

In this exercise we will trace the blood flow through various places in the body. As you trace, keep the following hints in mind:

- Don't forget about the hepatic portal system! Remember that most venous blood coming from the abdominal organs has to go through the hepatic portal vein before it can enter the general circulation.

- If you are going through the brain, don't leave out the dural sinuses.

- If you start in a vein, you have to go through the venous system and through the heart before you can get back to the arterial system.

- If you start in an artery, you have to go through the arterial system and then through a *capillary bed* before you can go through the venous system. You can't go backward through the arterial system—that's cheating!

- If you start in an artery and end in an artery, you likely will have to go through the arterial circuit, through a capillary bed, through the venous circuit, back to the heart and lungs, and *then* re-enter the arterial circuit. Whew!

- Sometimes there is more than one right answer, as one may take multiple paths.

- Below is an example, where we have started in the right popliteal vein, and ended in the left internal carotid artery:

  Start: right popliteal vein ➤ right femoral vein ➤ right external iliac vein ➤ right common iliac vein ➤ inferior vena cava ➤ right atrium ➤ tricuspid valve ➤ right ventricle ➤ pulmonic valve ➤ pulmonary artery ➤ lungs ➤ pulmonary veins ➤ left atrium ➤ mitral valve ➤ left ventricle ➤ aortic valve ➤ ascending aorta ➤ aortic arch ➤ left common carotid artery ➤ left internal carotid artery ➤ **End**

Wasn't that easy?

Procedure

Trace the path of blood flow through the following circuits, using the example above for reference. It is helpful to draw the pathway out on a laminated outline of the human body as you trace.

1. *Start*: Right radial vein
End: Right renal artery

2. *Start*: Left coronary artery
   End: Dorsalis pedis artery

3. *Start*: Superior mesenteric vein
   End: Superior mesenteric artery

4. *Start*: Renal artery
   End: Internal jugular vein

3 EXERCISE 3:
Histology of the Blood Vessel Wall _____

The blood vessel wall has three distinct layers:

1. *Tunica interna*: the innermost lining of the blood vessel, consisting of a specialized type of simple squamous epithelium called *endothelium*. It rests on top of a small amount of connective tissue.

2. *Tunica media*: the middle layer of the blood vessel wall, consisting of smooth muscle and elastic fibers. The smooth muscle, which is innervated by the sympathetic nervous system, controls the diameter of the vessel, playing an important role in tissue perfusion and blood pressure. The elastic fibers allow the vessel to expand with changing pressure and return to its original shape and diameter.

3. *Tunica externa*: the outer protective layer, consisting of fibrous connective tissue with abundant collagen fibers.

The characteristics of the three layers of the blood vessel wall are considerably different in arteries, capillaries, and veins. Following are some hints to help you distinguish between these three vessels:

● Arteries have a much thicker tunica media, with prominent elastic fibers that typically appear as a wavy purple line.

● Veins have a thin tunica media with few elastic fibers. Because the wall is so much thinner, the lumen is wider.

● Capillaries are extremely thin-walled and consist only of a thin tunica interna. The smallest capillaries are large enough for only one red blood cell to fit through at a time.

Examine prepared microscope slides of an artery, capillary, and vein (if available, compare an elastic artery to a muscular artery). Use colored pencils to draw what you see, and label your diagrams with the terms indicated.

1. Artery (see Figures 12.27 and 12.28 in your lab atlas)
   a. Tunica interna (endothelium)
   b. Tunica media
      (1) Smooth muscle
      (2) Elastic fibers
   c. Tunica externa
   d. Lumen

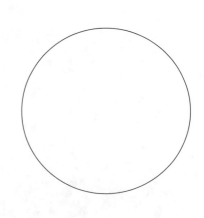

2. Capillary (See Figure 12.30 in your lab atlas.)
   a. Tunica interna
   b. Blood cell(s)

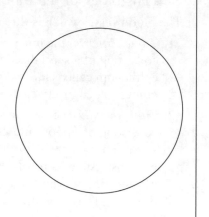

3. Vein (See Figure 12.27 in your lab atlas.)
   a. Tunica interna
   b. Tunica media
      (1) Smooth muscle
   c. Tunica externa
   d. Lumen

✔ **Check Your Understanding**

1. Why do arteries require more elastic fibers than veins?

2. Certain diseases, termed *collagen vascular diseases*, are characterized by having defects in the collagen in the tunica externa. What effects would you expect such diseases to have on the blood vessels?

18. Cardiovascular Physiology

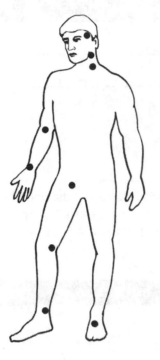

■ OBJECTIVES

Once you have completed this unit, you should be able to:

1. Auscultate the sounds of the heart.

2. Palpate and grade arterial pulses at various points.

3. Measure and record blood pressure using a stethoscope and sphygmomanometer.

4. Determine the effects of the autonomic nervous system on blood pressure and pulse rate.

5. Measure and record the ankle-brachial index using a Doppler ultrasound device.

■ MATERIALS

- Anatomical models: heart, torso
- Stethoscopes, with diaphragm and bell
- Sphygmomanometers
- Alcohol with cotton balls
- Buckets of ice water
- Doppler ultrasound
- Ultrasound gel

PRE-LAB EXERCISES

Prior to coming to lab, use your text and lab atlas to complete the following exercises.

PRE-LAB EXERCISE 1:
Heart Sounds _____

In this lab we will be performing a procedure known as *auscultation* of the heart sounds. Auscultation means "to examine by listening." To do this, we will use a stethoscope and listen for two distinct heart sounds, called S1 and S2. Both heart sounds are caused by the closing of valves in the heart during the cardiac cycle. Use your text to assist you in determining the properties of the heart sounds and recording them in the following chart.

HEART SOUND	CAUSE	TIMING IN THE CARDIAC CYCLE
S1		
S2		

PRE-LAB EXERCISE 2:
Autonomic Nervous System and the Cardiovascular System _____

As we discussed in Unit 14, the autonomic nervous system (ANS) is one of the most important systems in the body because of its profound effects on other body systems. The cardiovascular system is no exception. To better understand the results you will obtain from your experiments today, a review of the ANS is required. Fill in the following chart with the properties of the ANS.

ORGAN	SNS* EFFECTS	PSNS** EFFECTS
Heart		
Blood vessels		
Net effect on blood pressure		
Lungs (bronchial smooth muscle)		
Metabolism		
Urinary functions		
Digestive functions		

* SNS = sympathetic nervous system
** PSNS = parasympathetic nervous system

EXERCISES

A part of every physical examination includes an examination of the cardiovascular system, in which the pulse is taken, heart sounds are auscultated, and the blood pressure is measured. In this unit you will learn how to perform these common procedures, as well as the basis for their interpretation. You also will become familiar with the ankle-brachial index, a more specialized test of blood flow to the lower extremities.

EXERCISE 1:
Auscultation of Heart Sounds

In this exercise we will be using a stethoscope to auscultate heart sounds. Most stethoscopes contain the following parts (see Figure 18.1):

- *Earpieces*: these are gently inserted into the external auditory canal.
- *Diaphragm*: this is the broad, flat side of the end of the stethoscope. It is used to auscultate higher-pitched sounds and is the side used most often in auscultation of heart sounds.

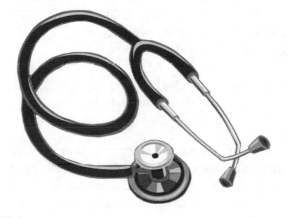

FIGURE 18.1
Diagram of a stethoscope.

- *Bell*: this is the concave, smaller side of the end of the stethoscope. It is used to auscultate lower-pitched sounds.

Note that sounds are not audible through both the bell and the diaphragm at the same time. Typically, the end can be flipped from one side to the next. Before auscultating with either side, lightly tap the end to ensure that you can hear sound through it. Note also that the ends of some stethoscopes have only one side—the diaphragm. In this case, placing light pressure on the end as you are auscultating yields sounds associated with the diaphragm while placing heavier pressure yields sounds associated with the bell.

Heart sounds typically are auscultated in five areas. The areas are named for the valve that is heard best at that specific location. Their positions are described relative to the sternum and the spaces between the ribs, known as *intercostal spaces*. The first is the space between the first and second rib, which is roughly below the clavicle. From the clavicle, you can count down to consecutive spaces, to auscultate in the appropriate areas (see Figure 18.2).

1. *Tricuspid area*: This is the area where the sounds of the tricuspid valve are best heard. It is located in the second intercostal space, at the right sternal border.

2. *Pulmonic area*: The pulmonic valve is best heard over this area, and it is in the second intercostal space, at the left sternal border.

3. *Aortic area*: The sounds produced by the aortic valve are best heard over this area, located in the fourth intercostal space, at the left sternal border.

4. *Mitral area*: The mitral valve is best auscultated over this region, which is located in the fifth intercostal space, at the midclavicular line (draw an imaginary line down the middle of the clavicle, roughly in line with the nipple).

While listening to heart sounds, several factors are checked:

- *Rate*: number of heartbeats per minute. If the rate is more than 100, it is termed *tachycardia*. If the rate is below 60 beats per minute, it is termed *bradycardia*.

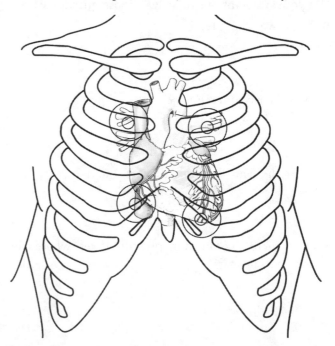

FIGURE 18.2
Thorax, showing areas of auscultation.

- *Regularity*: even rhythm of the heart, with no extra or skipped beats.
- *Additional heart sounds*: extra sounds in addition to the two heart sounds—S1, or the "lub," and S2, the "dub." If any additional beats are heard, it could be a sign of pathology. (Note that the presence of an additional heart sound does not necessarily denote pathology. Occasional splitting of the heart sounds is heard with ventilation and is called *physiological splitting of S2*.)
- *Murmurs*: clicking or "swooshing" noises heard between the heart sounds. These are caused by a valve leaking, called *regurgitation*, or by a valve that has lost its pliability, called *stenosis*.

Procedure: Auscultation

1. Obtain a stethoscope and clean the earpieces and diaphragm with alcohol and cotton swabs.

2. Place the earpieces in your ears and gently tap the diaphragm to ensure that it is on the proper side. If it is not, flip it to the other side.

3. Lightly place the diaphragm on your partner's chest in the tricuspid area.

> ➤ NOTE: YOU MAY WISH TO HAVE YOUR LAB PARTNER PLACE THE STETHOSCOPE ON HIS OR HER CHEST UNDER THE SHIRT, AS THE SOUNDS ARE HEARD BEST ON BARE SKIN.

4. Auscultate several cardiac cycles, determining which sound is S1, and which sound is S2.

5. Move on to the next area, and repeat.

6. Use your observations to answer these questions:
   - What was the rate?

   - Was the rhythm regular?

   - Were there any additional heart sounds?

   - Were there any murmurs present? If yes, where were they best heard?

2 EXERCISE 2: Palpation of Pulses

Feeling the pulse, a process known as *palpation*, at various superficial arterial points is a routine procedure performed in the medical setting. It is done to assess rate, rhythm, and regularity of the heartbeat, as well as to assess circulation to different parts of the body. (In this case, the pulses always are taken on both sides so they may be compared and any differences noted). The pulses commonly measured are those found at the radial, ulnar, brachial, carotid, temporal, femoral, popliteal, posterior tibial, and dorsalis pedis arteries. Figure 18.3 shows the pulse points.

FIGURE 18.3
Common pulse points.

When pulses are palpated, they are *graded* according to a standard scale. This allows for unambiguous communication between health care professionals, as well as for measuring the progress or deterioration of a patient's condition. The scale utilizes four grades:

Grade 0/4: The pulse is absent.
Grade 1/4: The pulse is barely or only lightly palpable.
Grade 2/4: The pulse is normal.
Grade 3/4: The pulse is quite strong.
Grade 4/4: The pulse is bounding, and visible through the skin.

Note that this scale has no negative numbers, nor are there any half-numbers (e.g., 1.5/4). In a healthy person most pulses will be graded as a 2/4, although the occasional pulse is weak or absent. This is simply normal anatomical variation and does not signify pathology. Students often mistakenly grade any strong pulse (such as the carotid pulse) as 4/4. If a pulse were truly 4/4, however, the patient would be symptomatic, as that is a sign of extremely high blood pressure in that artery.

Procedure: Pulse Palpation

1. Wash your hands prior to palpating your lab partner's pulses.

2. On a model or diagram, locate the artery you are palpating. Note that we will not be palpating each pulse illustrated in Figure 18.3.

3. Lightly place your index finger and middle finger over the artery. You may increase the pressure slightly, but be careful not to press too hard. You could cut off

circulation through the artery and also could mistake the pulse in your fingertips for your partner's pulse. If you are unsure if the pulse is yours or your partner's, feel the lab table. If the lab table "has a pulse," you are feeling the pulse in your own fingertips.

4. Palpate only one side (right or left) at a time, especially in the carotid artery. Palpation of both carotid arteries simultaneously could induce a phenomenon called the *baroreceptor reflex*, and your partner may lose consciousness.

5. Grade your pulses according to the scale above, and record this data in the chart below.

ARTERY	RIGHT SIDE GRADE	LEFT SIDE GRADE
Carotid		
Temporal		
Brachial		
Radial		
Ulnar		
Dorsalis pedis		
Posterior tibial		

✔ **Check Your Understanding**

1. What might it mean if a pulse were palpable, graded at 2/4, on the right limb, and absent, graded 0/4, on the left limb?

2. Palpation of both carotid arteries at the same time initiates a reflex known as the *baroreceptor reflex*, which results from pressure on the carotid sinus. This reflex can be used to our benefit in a procedure known as a *carotid sinus massage*. This procedure is performed when the blood pressure or heart rate is pathologically high. Why would a "massage" of the carotid sinus help to treat these conditions?

3 EXERCISE 3:
Blood Pressure _____

Blood pressure is determined by using an instrument called a *sphygmomanometer* and a stethoscope. This procedure yields two pressure readings:

1. *Systolic pressure*: the pressure in the arteries during ventricular systole and, naturally, the larger of the two readings.

2. *Diastolic pressure*: the pressure in the arteries during ventricular diastole, the smaller of the two readings.

These readings are taken using the sphygmomanometer to compress the brachial artery, which cuts off blood flow through that artery. The stethoscope is utilized to auscultate the brachial artery for sounds known as *sounds of Korotkoff*, which result from the turbulent blood flow resuming through the previously clamped-off artery.

Procedure: Blood Pressure

1. Obtain a stethoscope and sphygmomanometer of the appropriate size (about 80% of the circumference of the arm).

2. Clean the earpieces and diaphragm as in Exercise 1.

3. Wrap the cuff around your partner's arm. It should be about 1½ inches proximal to the antecubital fossa. Also, it should not be noticeably tight, but it should stay in place when you are not holding it.

4. Place the diaphragm of your stethoscope over the brachial artery. You should *not* hear anything at this point.

5. Support your partner's arm by cradling it in your arm, or have your partner rest his/her arm on the lab table.

6. Close the screw of the sphygmomanometer (by the bulb), and inflate the cuff by squeezing the bulb several times. Pay attention to the level of pressure you are applying by watching the pressure gauge. You should not inflate it beyond about 30 mmHg above your partner's normal systolic pressure (for most people, this is no higher than 180mmHg).

7. Slowly open the screw, watching the pressure gauge and listening to the brachial artery with your stethoscope.

8. Eventually you will see the needle on the pressure gauge begin to bounce; at about the same time, you will begin to hear the pulse in the brachial artery. The pressure at which this first happens is the *systolic pressure.*

9. Continue to listen and watch the gauge until you can no longer hear the pulse. At this point, the needle on the gauge will stop rhythmically bouncing. The pressure at which this happens is the *diastolic pressure.*

Practice reading 1: _____

Practice reading 2: _____

Effect of Autonomic Nervous System on Blood Pressure and Pulse

From the discussion of the nervous system, we know that the autonomic nervous system (ANS) exerts a great deal of control over blood pressure. In this exercise we will demonstrate how the blood pressure and pulse rate can change with activation of the sympathetic and parasympathetic nervous systems.

Blood pressure is determined by three factors:

1. *Cardiac output*: the amount of blood each ventricle pumps in one minute. It is determined by two factors: (1) stroke volume, or the amount pumped with each beat, and (2) heart rate.

2. *Peripheral resistance*: impedance to blood flow encountered in the system; determined largely by the degree of vasoconstriction or vasodilation in the systemic circulation. Vasoconstriction increases peripheral resistance, and resistance is decreased by vasodilation.

3. *Blood volume*: the amount of blood found in the blood vessels at any given time. It is greatly influenced by overall fluid volume, which is controlled by the kidneys.

Note that cardiac output and peripheral resistance are factors that can be altered quite quickly to change blood pressure, but alterations to blood volume occur

relatively slowly, generally taking two to three days to have a noticeable effect.

The ANS can influence each of the above three factors both directly and indirectly. This exercise, however, allows us to visualize only the effects on cardiac output and peripheral resistance (unless you want to stay in lab for the next two days—but I'm guessing you probably don't want to do that!).

Procedure

1. Measure your partner's blood pressure and pulse rate at rest (seated).

2. Have your partner immerse one hand into a bucket of ice water.

3. Repeat the blood pressure and pulse measurements with your partner's hand still in the ice water.

➤ NOTE: BE KIND TO YOUR PARTNER AND DO THIS QUICKLY!

4. Have your partner remove his/her hand from the ice water.

5. Wait 5 minutes, then repeat the blood pressure and pulse measurements. Record your results in the chart below.

TEST SITUATION	BLOOD PRESSURE	PULSE RATE
At rest		
After immersing in ice water		
5 minutes after removing from ice water		

✔ Check Your Understanding

1. Which situation(s) represented activation of the sympathetic nervous system?

2. What happened to the blood pressure and pulse rate in this (these) situation(s)? Why? (Which of the three factors were affected, and how?)

3. What situation(s) represented activation of the parasympathetic nervous system?

4. What happened to the blood pressure and pulse rate in this (these) situation(s)? Why?

4 EXERCISE 4: Determination of the Ankle-Brachial Index

Vascular disease, a relatively common diagnosis, can be caused by diabetes, atherosclerosis, and numerous other conditions. Typically, vascular disease affects the blood vessels of the legs rather than those of the arms, which has led to the development of a test known as the ankle-brachial index (ABI). The ABI compares the blood pressure in the legs (the "ankle" portion) to the pressure in the arms (the "brachial" portion).

Typically the ABI is a decimal number because ankle pressure is slightly lower than the brachial pressure. This is because the legs are more distant from the heart than the arms are, which causes a consequent decline in blood pressure. Note that in younger or more athletic patients, the value of the ABI is often greater than one. This is simply a result of more muscle mass in the lower limb, which is difficult to completely compress with a sphygmomanometer cuff.

Interpretation of the ABI is as follows:

0.9–1.0 = normal
0.5–0.9 = peripheral vascular disease
0.2–0.5 = intermittent claudication (temporary blockages to blood flow)
<0.2   = tissue death

The brachial pressure in this test is performed in the standard way, with a sphygmomanometer and a stethoscope. Because you cannot easily auscultate either of the pedal pulses, however, a Doppler ultrasound device is used to hear the blood flow. A Doppler ultrasound device uses soundwaves transmitted through a liquid medium (ultrasound gel) to produce audible sounds of the blood flow through the vessel.

Procedure: Ankle-Brachial Index

1. Wrap the sphygmomanometer cuff around the ankle.

2. Palpate for either the dorsalis pedis or posterior tibial pulses.

3. Place a small amount of ultrasound gel over the pulse, and place the Doppler probe in the gel, lightly touching the skin.

> ➤ NOTE: IF YOU PRESS TOO HARD WITH THE PROBE, IT IS POSSIBLE TO CUT OFF BLOOD FLOW, AND THEREFORE ALL SOUND.

4. Once you can hear sounds of the blood flowing, inflate the sphygmomanometer cuff until you no longer hear any sound. At this point, slowly deflate the cuff, making sure to hold the probe in place. When you first hear blood flowing again, this is the systolic ankle pressure. You will not obtain a diastolic pressure for the ankle.

Remember that we record the diastolic pressure as the number when we can no longer hear blood flowing, but with the Doppler, you will continue to hear blood flow; therefore, there will be no diastolic reading.

5. To calculate the ABI, record the systolic pressure of the ankle that you obtained in number 4 above, and the resting systolic pressure of the arm that you obtained earlier, in Exercise 3. Note that you will record *only* the systolic pressure, not the diastolic pressure. To get the ABI, simply divide the reading of the ankle by the reading of the arm. You most likely will get a decimal number, and this is your ABI.

Systolic pressure of the ankle: _____

Systolic pressure of the arm (brachial): _____

Ankle pressure / Brachial pressure: _____

19. Blood

■ OBJECTIVES

Once you have completed this unit, you should be able to:

1. Identify the formed elements of blood.

2. Perform blood typing of the ABO and Rh blood groups using simulated blood.

3. Explain the genetics of blood types.

■ MATERIALS

- Simulated blood types: A+, A−, B+, B−, AB+, AB−, O+, and O−
- Anti-A, anti-B, and anti-Rh antisera
- Spot plates
- Blood smear slides
- "Murder mystery game" with simulated blood of suspects, victims, rooms, and weapons

PRE-LAB EXERCISES

Prior to coming to lab, use your text and lab atlas to complete the following exercises.

1 PRE-LAB EXERCISE 1: Formed Elements _____

In this unit we will identify various formed elements on a peripheral blood smear. Each formed element has unique morphological characteristics, as well as unique functions.

Use your text and Figures 12.32 and 12.34 in your lab atlas to fill in the following chart pertaining to formed elements.

FORMED ELEMENT	NUCLEUS SHAPE	CYTOPLASM AND/OR GRANULE COLOR	FUNCTION	PREVALENCE
Erythrocyte				
Neutrophil				
Eosinophil				
Basophil				
Lymphocyte				
Monocyte				
Platelet				

2 PRE-LAB EXERCISE 2: The Genetics of Blood Typing _____

When discussing genetics, we refer to two properties of the cell:

1. *Phenotype* is the physical expression of a specific trait, such as hair color or blood type, based on genetic and environmental influences. Traits such as "blonde hair" and "brown eyes" refer to an individual's phenotype. To determine the phenotype of blood cells, we perform blood typing with antisera, which we will learn about in Exercise 2. The result of that blood typing procedure (for example, A+, O−, or AB−) is the phenotype.

2. *Genotype* is the genetic makeup, as distinguished from the physical appearance, of an organism. It is what the cell looks like genetically. If you were to perform DNA typing, the genotype is what you would be investigating. Remember that for all genes in the body, we have two copies, one from the maternal side and one from the paternal side. For blood, we cannot determine the precise genotype without doing DNA typing. But if we know the phenotype, we can determine *possible* genotypes. We know the following facts:

   a. From each parent, we can inherit one of the following ABO genes:

   (1) A gene: codes for A antigens

   (2) B gene: codes for B antigens

   (3) O gene: codes for neither A nor B antigen (please keep in mind that there is no "O" antigen)

   b. From each parent, we can inherit one of the following Rh factor genes:

   (1) Rh gene: codes for the Rh antigen, denoted as "+"

   (2) No Rh gene: does not code for the Rh antigen, denoted as "−"

We can put all of this information into a table to determine which phenotypes accompany which possible genotypes. Fill in the remainder of the table, following the pattern from the first two examples.

GENOTYPE(S)	PHENOTYPE
AA++, AA+−, AO++, AO+−	A+
AA−−, AO−−	A−
BB++, BB+−, BO++, BO+−	
	B−
OO++, OO+−	
	O−
	AB+
AB−−	

EXERCISES

Although safety concerns often preclude the use of real blood in the laboratory, we can still demonstrate important principles of blood by viewing prepared microscope slides of blood cells and by using simulated blood to demonstrate blood typing. When using simulated blood, there is no real concern over bloodborne diseases, but do keep in mind that the simulated blood contains chemicals that may be hazardous. As such, use all appropriate safety protocols when handling the simulated blood.

In this unit, you will identify the formed elements of blood on microscope slides, and you will play a murder mystery game in which you use simulated blood to apply blood typing techniques.

▌EXERCISE 1:
Formed Elements (Cells) of Blood _____

Formed elements, which make up the cellular portion, or "living matrix," of blood, account for about 45% of the volume of whole blood. Each class of formed element has a distinctive appearance, some with uniquely colored cytoplasmic granules, and all with a characteristic shape of the nucleus. Formed elements can be divided into three classes:

1. *Erythrocytes* (red blood cells): have the function of carrying oxygen around the body on an iron-containing molecule called hemoglobin.

2. *Leukocytes* (white blood cells): play a role in the immune system. The two subclasses of leukocytes are:

   a. *Granulocytes*: cells containing visible cytoplasmic granules, including neutrophils, eosinophils, and basophils.

   b. *Agranulocytes*: cells lacking cytoplasmic granules, including lymphocytes and monocytes.

3. *Platelets*: cellular fragments involved in blood clotting.

All of the formed elements have unique morphological characteristics, which you learned in the Pre-Lab Exercises. These features can be utilized to view individual cells on a blood slide called a *peripheral blood smear*. For this exercise, obtain a peripheral blood smear slide, examine it on high power, and scroll through it attempting to find each of the formed elements. Note that you may have to find a second slide to locate certain cells, because some types, such as

basophils and eosinophils, are rare. In the spaces below, use colored pencils to draw and describe each formed element that you locate. Refer to Figure 12.34 in your atlas for reference.

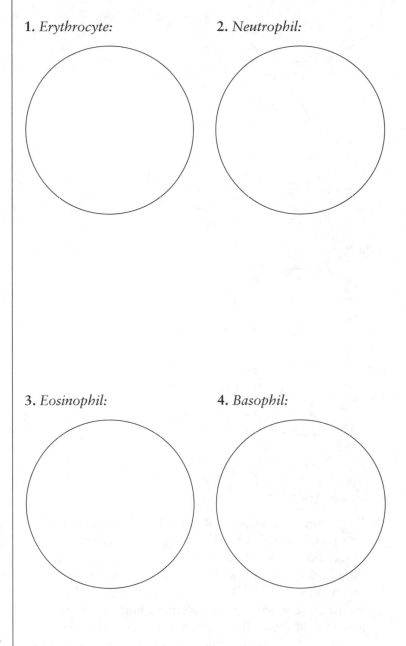

1. *Erythrocyte:*

2. *Neutrophil:*

3. *Eosinophil:*

4. *Basophil:*

5. *Monocyte:*   6. *Lymphocyte:*

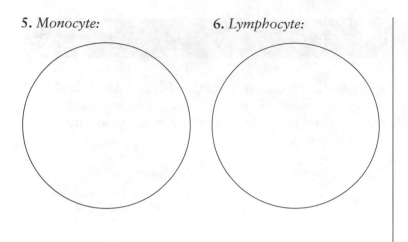

7. *Platelets:*

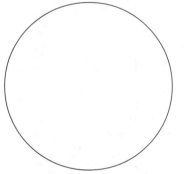

✔ Check Your Understanding

1. When someone is admitted to the hospital, one of the first procedures that health professionals perform is a blood draw. One lab value that is checked is a white blood cell (WBC) count. If the WBC count is high (normal range = $4,000 - 11,000/mm^3$), this could indicate the presence of an infection. Further analysis (called a differential) is performed to determine the relative prevalence of different types of white blood cells. What type of infection do you think would be present if:

   a. Neutrophils were elevated? Explain your reasoning.

   b. Lymphocytes were elevated? Explain your reasoning.

   c. Eosinophils were elevated? Explain your reasoning.

2. Another lab value that is monitored routinely is the total red blood cell (RBC) count. A normal RBC count is typically $4.2 - 5.9$ million/mm^3. Having a low RBC count is called *anemia*. Considering the function of RBCs, predict the possible effects of having anemia.

2 EXERCISE 2: ABO and Rh Blood Groups _____

Blood typing is done by checking the blood for the presence or absence of specific surface markers called antigens.

- Type A blood has A antigens on the cell surface.
- Type B blood has B antigens on the cell surface.
- Type AB blood has both A and B antigens on the cell surface.
- Type O blood has neither A nor B antigens on the cell surface.

An additional antigen that may be present on the red blood cell is the Rh antigen.

- Blood that has the Rh antigen is denoted as Rh positive (e.g., A +).

- Blood that lacks the Rh antigen is denoted as Rh negative (e.g., A−).

To determine the type of antigens that are present, the blood is combined with different antisera. *Antisera* are solutions that contain antibodies that bind with specific antigens. If there is a reaction, it typically is represented by the clumping or agglutination of the blood sample. In this exercise we are using simulated blood, and a positive reaction is denoted by a color change.

- Anti-A antiserum contains anti-A antibodies and reacts with blood with A antigens.
- Anti-B antiserum contains anti-B antibodies and reacts with blood with B antigens.
- Anti-Rh antiserum contains anti-Rh antibodies and reacts with blood with Rh antigens.

Antigen–Antibody Reactions

> ➤ NOTE: GLOVES AND SAFETY GLASSES ARE REQUIRED.

This exercise allows you to examine the antigen–antibody reactions of known blood types. Each table should take one spot plate and one set of dropper bottles. The bottles are labeled A+, AB−, B+, and O− to represent each of those blood types, and anti-A, anti-B, and anti-Rh to represent the different antisera.

Procedure

Use the diagram below as a guide to placement of samples in the wells.

1. Label wells on the spot plate as wells 1–12.

2. Drop two drops of type A+ blood in well 1, well 2, and well 3.

3. Drop two drops of type B− blood in well 4, well 5, and well 6.

4. Drop two drops of type AB+ blood in well 7, well 8, and well 9.

5. Drop two drops of type O− blood in well 10, well 11, and well 12.

6. Add two drops of the anti-A antiserum to wells 1, 4, 7, and 10.

7. Add two drops of the anti-B antiserum to wells 2, 5, 8, and 11.

8. Add two drops of the anti-Rh antiserum to wells 3, 6, 9, and 12.

9. Observe the samples for changes in color symbolizing the agglutination or clumping that would normally occur

between antisera and specific blood types. Use your results to complete the chart below.

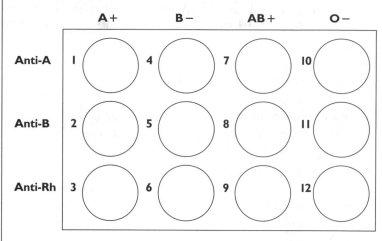

Fill in your results in the chart below.

BLOOD TYPE	REACTED WITH ANTI-A? (YES/NO)	REACTED WITH ANTI-B? (YES/NO)	REACTED WITH ANTI-RH? (YES/NO)	ANTIGENS PRESENT ON CELL SURFACE
A+				
B−				
AB+				
O−				

3 EXERCISE 3: *Blood Typing Mystery* Game _____

In this game you will be applying the blood typing techniques that you learned in Exercise 2. In addition, you will be learning about and applying the genetics of blood typing. Refer to the Pre-Lab Exercises to review the genetics of blood typing.

Each of the following cases presents a victim, a murderer, three suspects, three possible murder rooms, and three possible murder weapons. Your job, as the detective, is to determine the identity of the murderer, which weapon he or she used, and in which room the crime was committed. In addition, you will need to sort out some sticky paternity issues!

Procedure

Basic Instructions

1. For each case there is a unique set of bottles marked with a number that corresponds to the specific cases (i.e., the bottles are marked 1 for Case 1, 2 for Case 2, and 3 for Case 3). The murderer, rooms, and murder weapons are different for each case, but the cast of characters remains the same.

2. To begin the game, assemble into groups of two or three students each and obtain a spot plate for each group. Choose one set of samples to test (e.g., the rooms from Case 1, the suspects from Case 2, or the weapons from Case 3).

3. Test the samples by placing two drops of the sample in three separate wells. Add two drops of anti-A to the first well, add two drops of anti-B to the second well, and add two drops of anti-Rh to the third well. When you have tested each of the samples, *return them to their proper places* in the front of the lab.

4. Read and record the blood type by watching for a reaction with the antisera. Remember that this is simulated blood, just like in Exercise 2. A positive reaction is denoted by a color change.

5. To determine the:
   - *Murderer*: Match the blood type of one of the *suspects* to that of the murderer.
   - *Weapon and room*: Match the blood type of the *victim* to the blood types found in the rooms and on the weapon.

Case 1: Ms. Magenta

We enter the scene to find the dearly departed Ms. Magenta. Forensic analysis determines that there are two types of blood on the body. One blood type is Ms. Magenta's, and the other blood type is a trace amount left behind from the murderer.

Ms. Magenta's blood type: _____

Murderer's blood type: _____

We have three suspects:

1. *Mrs. Blanc* was being blackmailed by Ms. Magenta and Col. Lemon. They had discovered that she was *really* a man. Mrs. Blanc knew this fact would ruin her reputation at the country club.

2. *Col. Lemon* wanted to keep the blackmail money for himself, and wanted Ms. Magenta out of the way.

3. *Mr. Olive* was having an affair with Ms. Magenta, and she was threatening to tell his wife.

Mrs. Blanc's blood type: _____

Col. Lemon's blood type: _____

Mr. Olive's blood type: _____

We have three possible murder rooms:

Ballroom blood type: _____

Library blood type: _____

Den blood type: _____

We have three possible murder weapons:

Candlestick blood type: _____

Noose blood type: _____

Knife blood type: _____

Case 1 Conclusion:

Ms. Magenta was killed by _____, in the _____, with the _____.

Who's Your Daddy? Part 1

The investigators have discovered that Ms. Magenta has a 20-year-old daughter, Ms. Rose, about whom nobody knew. Given Ms. Magenta's propensity for affairs, it has become difficult to determine the paternity of Ms. Rose. The three possible fathers are: Col. Lemon, Mr. Olive, and Professor Purple. You are asked to determine if any of these three men may be ruled out as Ms. Rose's father on the basis of their blood types. Follow the same procedure as you did in Case 1 to type these blood samples.

1. What is Ms. Magenta's phenotype? Possible genotypes?

2. What is Ms. Rose's phenotype? Possible genotypes?

3. What are the phenotypes of the three potential fathers? Their possible genotypes?

4. Given this information, can any of these men be excluded? Why or why not?

Case 2: Col. Lemon

Our next victim is poor Col. Lemon. On his body, we find his blood, and also trace amounts of the blood of another person, presumably the murderer.

Colonel Lemon's blood type: _____

Murderer's blood type: _____

We have three potential suspects:

1. *Mrs. Blanc.* Now, with Ms. Magenta out of the way, Mrs. Blanc could easily rid herself of her problem by disposing of the only other person who knows her secret—Col. Lemon.

2. *Professor Purple* was terribly angry with Col. Lemon, because the professor believed the Colonel stole his groundbreaking research into collagen lip injections.

3. *Mr. Olive* couldn't stand the Colonel because he thought he was having an affair with his wife, Mrs. Feather.

Mrs. Blanc's blood type: _____

Professor Purple's blood type: _____

Mr. Olive's blood type: _____

We have blood in three different rooms:

Hall blood type: _____

Kitchen blood type: _____

Billiards blood type: _____

Forensics found blood on three different weapons:

Copper pipe blood type: _____

Hammer blood type: _____

Revolver blood type: _____

Case 2 Conclusion:

Colonel Lemon was killed by _____,
in the _____, with the _____.

Case 3: Mr. Olive

Our next (and hopefully last) victim is Mr. Olive. Analysis demonstrates two blood types; one belonging to Mr. Olive, and trace amounts of another belonging to the murderer.

Mr. Olive's blood type: _____

Murderer's blood type: _____

We have three potential suspects:

1. *Mrs. Feather* found out that her husband, Mr. Olive, was having an affair and was quite angry.

2. *Mrs. Blanc* had a crush on Mrs. Feather and thought Mr. Olive didn't treat her well.

3. *Professor Purple* discovered that Mr. Olive—not Col. Lemon—had actually stolen the collagen lip implant research. Whoops!

Mrs. Feather's blood type: _____

Mrs. Blanc's blood type: _____

Professor Purple's blood type: _____

Blood was found in three rooms:

Lounge blood type: _____

Dining room blood type: _____

Greenhouse blood type: _____

We have three potential murder weapons:

Noose blood type: _____

Hammer blood type: _____

Revolver blood type: _____

Case 3 Conclusion:

Mr. Olive was killed by _____, in the
_____, with the _____.

Who's Your Daddy? Part 2

Gasp! It is now revealed that, because of a freak baby-switching incident, Ms. Magenta is in fact not Ms. Rose's mother, but instead it is Mrs. Feather! We have four possible fathers: Col. Lemon, Mrs. Blanc (remember—she's really a man), Professor Purple, and Mr. Olive.

1. What is Mrs. Feather's phenotype? Possible genotypes?

2. What is Ms. Rose's phenotype? Possible genotypes?

3. What are the phenotypes of the four potential fathers? Their possible genotypes?

4. Given this information, can any of the potential fathers be excluded? Why or why not?

5. What if Ms. Rose's blood type was O + ? How would this change the scenario, both *maternally* and *paternally*?

20. Lymphatics and Immunity

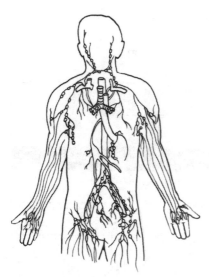

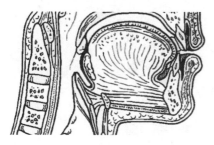

■ OBJECTIVES

Once you have completed this unit, you should be able to:

1. Identify structures of the lymphatic system on anatomical models and preserved specimens.

2. Describe the role of the immune system in processing lymphatic fluid as it is returned to the heart.

3. Trace the pathway of lymphatic fluid as it is returned to the heart.

4. Observe histological features of lymphoid organs on microscope slides.

■ MATERIALS

- Anatomical models: human torsos, intestinal villi, head and neck models
- Microscope slides: spleen, lymph node, Peyer's patch (ileum)
- Laminated outline of the human body
- Water-soluble marking pens
- Fetal pigs (or other preserved small mammal)
- Dissecting equipment
- Colored pencils

PRE-LAB EXERCISES

Prior to coming to lab, complete the following exercises, using Chapter 13 in your lab atlas and your textbook for reference.

1 PRE-LAB EXERCISE 1: Structures of Lymphatic System _____

Use your colored pencils to label and color-code Figure 20.1 with the structures listed in Exercise 1. See Figure 13.1 in your lab atlas for reference. Please note that all structures may not be visible on the diagrams.

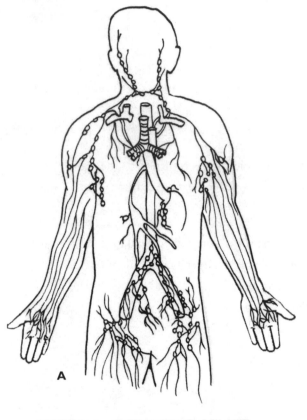

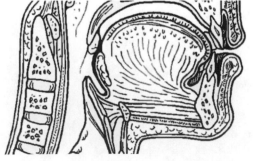

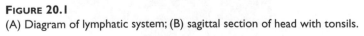

FIGURE 20.1
(A) Diagram of lymphatic system; (B) sagittal section of head with tonsils.

2 PRE-LAB EXERCISE 2: Lymphatic System Vessels _____

On the chart below, define each of the following vessels of the lymphatic system, noting which specific areas of the body that each vessel drains.

LYMPHATIC TRUNK	AREA(S) OF THE BODY DRAINED
Bronchomediastinal trunk	
Cervical trunk	
Axillary trunk	
Inguinal trunk	
Intestinal trunk	
LYMPHATIC DUCT	**LYMPHATIC TRUNKS DRAINED**
Right lymphatic duct	
Thoracic duct	

EXERCISES

The lymphatic system serves numerous homeostatic functions in the body. For one, it serves protective and defensive functions, combating harmful agents in the internal and external environments. It also works with the cardiovascular system to maintain fluid balance in the interstitium, and with the gastrointestinal system to absorb fats.

The exercises in this unit will introduce you to this diverse system. In the first exercise, you will identify the structures of the lymphatic system and trace the flow of lymph throughout the body. In the second exercise, you will examine three types of lymph tissue on microscope slides.

EXERCISE 1:
Lymphatic System Anatomy

The lymphatic system consists of a diverse group of organs that have three primary functions:

1. *Transport of excess interstitial fluid back to the heart.* In the cardiovascular system, hydrostatic pressure, the force of blood on the blood vessel wall, is stronger than colloid osmotic pressure, the force of proteins in the blood. This creates a gradient to push fluid out of the capillary beds and into the interstitial space. Approximately 1.5 ml/min of fluid is lost out of the circulation in this manner. This may not sound like a lot, but if this fluid were not returned to the blood vessels, we could lose our entire plasma volume in about 1 day! Luckily, the lymphatic system picks up this excess fluid and, through a series of lymph vessels, returns it to the cardiovascular system at the junction of the internal jugular and subclavian veins. The fluid is picked up first by small lymph capillaries that surround capillary beds. These capillaries are distinct from blood capillaries and contain highly permeable walls. From the lymph capillaries, the fluid, now called *lymph*, is delivered to larger lymph-collecting vessels, and finally to lymph trunks and lymph ducts before being returned to the circulation.

2. *Activation of immune system.* Several of the lymphatic organs function to activate the immune system. These include the *thymus*, an organ in which T lymphocytes mature, the *spleen*, which houses phagocytes, and the *tonsils*, aggregates of unencapsulated lymphoid tissue found in the oropharynx and nasopharynx. In addition, lymphoid organs called lymph nodes are found along the lymphatic vessels. The vessels deliver lymph into the nodes, where it is filtered so it can trap pathogens, toxins, and cells (such as cancer or virus-infected cells).

3. *Absorption of dietary fats.* Fats are not absorbed from the small intestine directly into the blood stream. Instead, fats enter a lymphatic vessel called a *lacteal*, after which they travel with the lymph to be deposited in the blood at the junction of the internal jugular and subclavian veins.

Identify the following structures of the lymphatic system on anatomical models and charts. Use the Pre-Lab Exercises and Chapter 13 in your lab atlas for reference.

Anatomical Structures of the Lymphatic System
(See Figures 13.1, 13.4, 13.6, and 13.10 in your lab atlas for reference.)

1. Lymph vessels
   a. Thoracic duct
   b. Right lymphatic duct
   c. Bronchomediastinal trunk
   d. Cervical (jugular) trunk
   e. Axillary trunk
   f. Inguinal trunk
   g. Intestinal trunk
   h. Lacteal
   i. Chyli cisterna
2. Lymph nodes
   a. Cervical lymph nodes
   b. Axillary lymph nodes
   c. Inguinal lymph nodes
   d. Mesenteric lymph nodes
3. Spleen
4. Thymus (this is best viewed on a fetal pig)
5. Mucosal-associated lymphoid tissue
6. Vermiform appendix
7. Tonsils
   a. Palatine tonsil
   b. Pharyngeal tonsil
   c. Lingual tonsil

Model Inventory

As you view the anatomical models and torsos in lab, list them on the inventory below, and state which of the preceding structures you are able to locate on each model.

MODEL/DIAGRAM	STRUCTURES IDENTIFIED

Fetal Pig Structures

> ➤ NOTE: SAFETY GLASSES AND GLOVES ARE REQUIRED.

The thymus degenerates in adults, so it often is not represented on anatomical models and torsos. Therefore, it is advantageous to view the thymus and other lymphatic organs in a preserved specimen as well. Fetal pigs are particularly well-suited to this task, as they have a prominent thymus.

Open a fetal pig if one is not already opened in the manner described in Chapter 20 of your lab atlas. Identify the following organs and describe their locations (see Figures 20.12 and 20.14 in your atlas for reference):

1. Spleen

   Description:

2. Thymus:

   Description:

Tracing the Flow of Lymph Through the Body

Trace the pathway of lymph flow from the starting point (given below) to the point at which the lymph fluid is returned to the cardiovascular system. Trace the flow through the major lymph-collecting vessels, trunks, and ducts, highlighting clusters of lymph nodes through which the lymph passes as it travels. First write the sequence of the flow, and then use differently colored water-soluble markers to draw the pathway on a laminated outline of the human body. If no outline is available, use Figure 20.2.

Trace the flow from the following locations:

1. Right foot

2. Left arm

3. Right cervical region

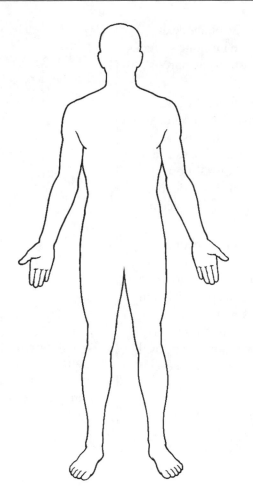

FIGURE 20.2
Outline of human body, anterior view.

✔ **Check Your Understanding**

1. What are some potential consequences of removing the spleen?

2. In a condition called DiGeorge Syndrome, infants are born with either an absent thymus or a thymus that isn't functional. What would the repercussions be from this disease?

3. Explain why blockage or removal of the lymphatic vessels can result in significant edema.

2 EXERCISE 2:
Histology of Lymphatic Organs _____

In this exercise, you will examine three different types of lymph tissue:

1. *Spleen:* The spleen consists of two types of tissue: (1) *red pulp*, which is involved in the destruction of red blood

cells, and (2) *white pulp*, which contains phagocytes and lymphocytes and plays a role in the immune system. On microscope slides, the white pulp stains as purplish nodules within the red pulp.

2. *Lymph nodes:* Lymph nodes are surrounded by a connective tissue capsule. The nodes consist of an outer cortex, which contains spherical clusters of cells (primarily B lymphocytes) called lymphatic nodules. The area of the cortex deep to the nodules houses mostly T lymphocytes. The innermost region of the node, the medulla, houses macrophages, which are highly active phagocytes.

3. *Peyer's patch:* A Peyer's patch is a cluster of incompletely encapsulated lymphoid tissue found in the ileum (the last portion of the small intestine). Examine the section on low power, and look for the mucosa, the epithelial lining of the ileum. As you scroll downward on the slide, you will enter the submucosa, connective tissue that stains mostly pink. In the submucosa you will note large, oval or teardrop-shaped, purplish clusters. These are Peyer's patches.

Examine prepared slides of the spleen, a lymph node, and the ileum. Use colored pencils to draw what you see, and label your drawings with the terms indicated.

Spleen
(See Figure 13.5 in your lab atlas.)

1. Red pulp
2. White pulp

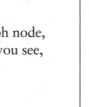

Lymph Node
(See Figure 13.9 in your lab atlas.)

1. Capsule
2. Cortex
3. Medulla
4. Germinal center (lymphatic nodule)

Ileum

1. Mucosa (simple columnar epithelial tissue)
2. Submucosa
3. Peyer's patch

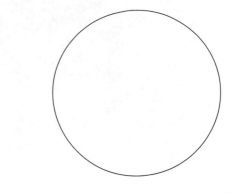

✔ Check Your Understanding

1. A common symptom of pharyngitis is swelling of the anterior cervical lymph nodes. Why would the lymph nodes swell in the presence of an infection? What else may cause swollen lymph nodes?

2. Why do you think the terminal portion of the small intestine would need abundant lymphatic tissue such as Peyer's patches? *(Hint: What colonizes the large intestine?)*

21. Respiratory System Anatomy

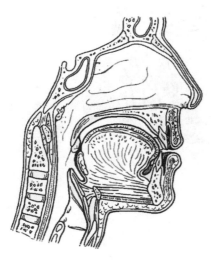

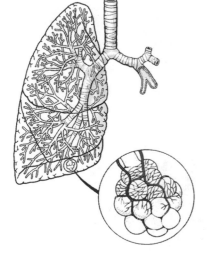

■ **OBJECTIVES**

Once you have completed this unit, you should be able to:

1. Identify structures of the respiratory system on models and fresh specimens.
2. Identify histologic features of different regions of the respiratory tract.
3. Examine the anatomical changes associated with inflation of the lungs of fresh or preserved specimens.

■ **MATERIALS**

● Anatomical models: lungs, bronchial tree, alveolar sac
● Microscope slides: trachea, alveoli
● Colored pencils
● Fetal pigs and dissection equipment
● Drinking straws
● Rubber hose
● Fresh lung specimens

PRE-LAB EXERCISES

Prior to coming to lab, use your text and lab atlas to complete the following exercises.

▌ PRE-LAB EXERCISE 1: Respiratory Anatomy _____

Figure 21.1 presents five diagrams of the respiratory system. Using the terms from Exercise 1, label these diagrams, then color-code them using colored pencils. See Figures 14.2–14.5 in your lab atlas for reference.

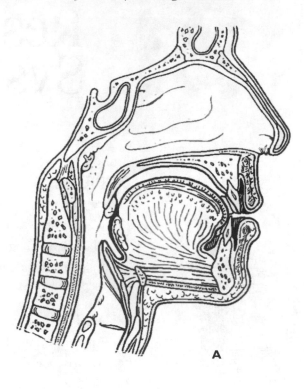

A

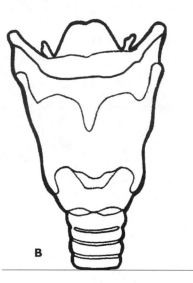

B

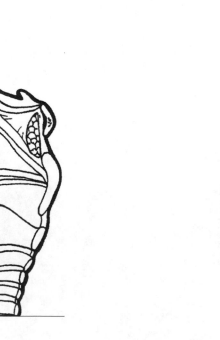

C

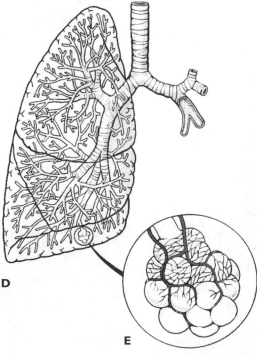

D

E

FIGURE 21.1

Diagrams of respiratory system: (A) sagitally sectioned head, (B) anterior view of larynx and trachea, (C) sagittal section of larynx, (D) trachea, bronchi, and lungs, (E) bronchiole with alveoli.

EXERCISES

As cells metabolize glucose to make ATP, they require oxygen and produce carbon dioxide. The respiratory system works closely with the cardiovascular system to perform the vital function of supplying the body with oxygen and ridding it of excess carbon dioxide through the process of *ventilation*.

The exercises in this unit will familiarize you with the structures of the respiratory system (Exercise 1), and the histology of the respiratory tract (Exercise 2). Finally, in Exercise 3 you will follow procedures to inflate preserved and fresh lung specimens.

EXERCISE 1:
Respiratory System Anatomy _____

The lungs are the principal organs of the respiratory system. They are composed of branching respiratory passages, terminal air sacs called *alveoli*, and elastic connective tissue. They are surrounded by double-layered serous membranes similar to the pericardial membranes—the *pleural membranes*:

1. *Parietal pleurae* cover the interior of the thoracic cavity and the superior surface of the diaphragm.

2. *Visceral pleurae* are an extension of the parietal pleurae that cover the lungs.

Between the two pleurae is a space called the *pleural cavity*, which is filled with serous fluid to reduce friction as the lungs expand and collapse.

The remaining respiratory organs are called, collectively, the *respiratory tract*. The respiratory tract may be divided into two regions or *zones*, according to function.

1. *The conducting zone* consists of most of the organs of the respiratory tract, and carries, or "conducts," air to the lower passages, where gas exchange takes place.

2. *The respiratory zone* is the region located in the terminal airways in which gas exchange takes place.

The conducting zone begins with the *nasal cavity* and the *paranasal sinuses*, designed to filter, warm, and humidify the inhaled air, which must be equal to 100% humidity and body temperature by the time it reaches the distal passages. The conducting zone continues with the *pharynx* (the throat —pronounced "fair-inks"), which has three divisions.

1. The *nasopharynx* is the region posterior to the nasal cavity, which should allow only air to pass through. The muscles of the soft palate move superiorly to close off the nasopharynx during swallowing to prevent food from entering the passage. Sometimes this mechanism fails (such as when a person is laughing and swallowing simultaneously), and the unfortunate result is that food or liquid comes out of the nose.

2. The *oropharynx* is the region posterior to the oral cavity, which holds both food and air.

3. The *laryngopharynx* is the intermediate region between the larynx and the esophagus.

Air passes from the pharynx to the *larynx* (voice box, pronounced "lair-inks"), which is a passage framed by nine cartilages (see Figures 14.3 and 14.4 in your lab atlas), including a flap of elastic cartilage called the *epiglottis*. The epiglottis covers the larynx during swallowing to prevent food from entering the larynx.

As its common name "voice box" implies, the larynx is the structure where sound is produced. It contains two sets of elastic ligaments, the *vocal folds*. The superior set of vocal folds, the *false vocal cords* (also called the vestibular folds), play no role in sound production. They do, however, serve an important sphincter function and can constrict to close off the larynx. The inferior set of vocal folds, called the *true vocal cords*, vibrate as air passes over them to produce sound.

Air next passes into a tube supported by C-shaped rings of hyaline cartilage called the *trachea*. In the mediastinum, the trachea bifurcates into two *primary bronchi*, which begin the bronchial tree. The right primary bronchus is shorter, straighter, and wider, and the left primary bronchus is longer, more horizontal, and narrower.

Each primary bronchus divides into smaller *secondary bronchi*, each of which serves one lobe of the lung. The secondary bronchi continue to branch, becoming tertiary bronchi, quaternary bronchi, and so on. By the time the passages are less than 1 millimeter in diameter, they are termed *bronchioles*.

At the *terminal bronchioles*, passageways smaller than 0.5 mm in diameter, the conducting zone ends. The respiratory zone begins with smaller branches called *respiratory bronchioles*, which have thin-walled sacs called alveoli in their walls. As the respiratory bronchioles progressively branch, the number of alveoli in the walls increases until the wall is made up exclusively of alveoli, at which point it is termed an *alveolar duct*. The terminal portions of the respiratory zone are called *alveolar sacs*, grape-like clusters of alveoli surrounded by *pulmonary capillaries*.

Identify the following structures of the respiratory system on models and diagrams, using your textbook and Figures 14.2–14.6 in your lab atlas as a guide.

Respiratory Structures

1. Lungs
   a. Hilus
   b. Parietal pleura
   c. Visceral pleura
   d. Pleural cavity
2. Right lung
   a. Upper, middle, lower lobes
   b. Horizontal fissure
   c. Oblique fissure
3. Left lung
   a. Upper and lower lobes
   b. Oblique fissure
   c. Cardiac notch
4. Nasal cavity
   a. Nares
   b. Vestibule
   c. Nasal septum
   d. Nasal conchae
      (1) Superior nasal conchae
      (2) Middle nasal conchae
      (3) Inferior nasal conchae
5. Paranasal sinuses
   a. Sphenoid sinus
   b. Ethmoid sinus
   c. Frontal sinus
   d. Maxillary sinus
6. Hard palate
7. Soft palate and uvula
8. Pharynx
   a. Nasopharynx
   b. Oropharynx
   c. Laryngopharynx
9. Larynx
   a. Epiglottis
   b. Thyroid cartilage
   c. Cricoid cartilage
   d. Cricothyroid ligament
   e. Corniculate cartilage
   f. Arytenoid cartilage
   g. Cuneiform cartilage
   h. False vocal cords
   i. True vocal cords
10. Trachea
    a. Hyaline cartilage rings
    b. Trachealis muscle
    c. Carina

11. Bronchi
    a. Right and left primary bronchi
    b. Secondary bronchi
12. Bronchioles
    a. Terminal bronchioles
    b. Respiratory bronchioles
    c. Alveolar duct
13. Alveoli and alveolar sacs
14. Muscles
    a. Diaphragm
    b. External intercostals
    c. Internal intercostals
15. Vascular structures
    a. Pulmonary arteries
    b. Pulmonary arterioles
    c. Pulmonary capillaries
    d. Pulmonary venules
    e. Pulmonary veins

Model Inventory

As you examine the anatomical models and diagrams in lab, list them on the inventory below and state which structures you are able to locate on each model.

MODEL/DIAGRAM	STRUCTURES IDENTIFIED

✔ Check Your Understanding

1. Note the C-shape of the hyaline cartilage rings of the trachea. Why do you think the rings are shaped this way, rather than being "O"-shaped? (*Hint*: Think about the structure behind the trachea.)

2. Conditions such as pneumonia and lung cancer can result in what is known as a *pleural effusion*, in which the pleural cavity becomes filled with an abnormal amount of fluid. What effects do you think a pleural effusion would have on ventilation? Explain.

2 EXERCISE 2: Histology of the Respiratory Tract_____

In this exercise we will examine different regions of the respiratory tract. In general, the respiratory tract consists of three histologic layers (see Figure 14.1 in your lab atlas):

1. *Mucosa*: Any passageway that opens to the outside of the body is lined with a protective mucosa. Throughout much of the respiratory tract, you will note that there are copious mucus-secreting goblet cells. The mucosa of the respiratory tract consists of epithelial tissue that is specialized for the specific region in which the tissue is found. Some examples are:

   a. *Nasopharynx*: pseudostratified ciliated columnar epithelium (sometimes called *respiratory epithelium*)

   b *Oropharynx and laryngopharynx*: stratified squamous epithelium

   c. *Larynx* (inferior to the vocal folds), trachea, and bronchi: pseudostratified ciliated columnar epithelium

   d. *Alveoli*: simple squamous epithelium

2. *Submucosa*: This is a layer of connective tissue that lies underneath the mucosa. It contains specialized seromucous glands that secrete watery mucus.

3. *Adventitia*: This outer layer consists of connective tissue and, in much of the respiratory tract, contains hyaline cartilage for support and protection. It also contains varying amounts of smooth muscle, particularly in the lower respiratory passages.

As you progress through different regions of the respiratory tract, you will note a few trends:

● The epithelial tissue changes from taller (pseudostratified) in the upper passages to shorter (cuboidal, then squamous) in the lower passages.

● The amount of hyaline cartilage gradually decreases (it is absent in bronchioles), as does the number of goblet cells.

● The amount of smooth muscle and elastic fibers increases.

Slides

Examine the following prepared microscope slides. Use colored pencils to draw what you see, and label the structures indicated below. Note that for most of the slides, it is useful to begin on low power to gain a full view of all histologic layers, then move up to higher power to identify specific structures on the slide.

Trachea

(See Figures 3.8 and 14.1 in your lab atlas.)

Label the following on your drawing:

1. Pseudostratified columnar epithelium

2. Cilia

3. Goblet cells

4. Submucosa

5. Seromucous glands

6. Hyaline cartilage

Bronchioles

(See Figure 14.7 in your lab atlas.)

Label the following on your drawing:

1. Simple columnar or cuboidal epithelium

2. Smooth muscle

3. Pulmonary arteriole

Alveoli

(See Figures 14.9–14.10 in your lab atlas.)

Label the following on your drawing:

1. Simple squamous epithelium

2. Respiratory bronchiole

3. Alveolar duct

4. Alveolar sac

✔ Check Your Understanding

1. Explain how the epithelium is adapted in each region of the respiratory tract so that its form follows its function.

2. Why do you think the upper respiratory passages have so many goblet cells and the lower respiratory passages have fewer?

3 EXERCISE 3: Lung Inflation

In this exercise you will examine fresh or preserved lung specimens and perform procedures to inflate the lungs.

Part 1: Performing a Cricothyroidotomy

A popular scene in movies shows a choking victim being rescued via a hole poked in the victim's neck (usually with a pocket knife), and a straw (or pen) inserted to restore an airway and save the victim's life. This produces a great dramatic effect and is in fact a legitimate procedure, called a *cricothyroidotomy*. This procedure is useful any time the upper airway is blocked, such as in choking, which is actually created by the resulting laryngospasm, not the inhaled object itself.

The procedure is performed by placing an incision (with a scalpel, hopefully, and not a pocket knife) in the cricothyroid ligament, which is a soft spot palpable between the thyroid and cricoid cartilages (see Figure 21.1, and also Figure 14.4 in your lab atlas). The airway then is reopened by inserting a tube through the incision and ventilating the patient.

In the following procedure you will perform a cricothyroidotomy on a fetal pig (or other preserved small mammal) and inflate the lungs.

Procedure

➤ NOTE: SAFETY GLASSES AND GLOVES ARE REQUIRED.

1. Obtain a fetal pig and dissection equipment, including a small scalpel.

2. Carefully expose the neck of the fetal pig, clearing away soft tissue so the larynx is clearly visible.

3. Locate the cricothyroid ligament and cut a small incision in its anterior surface.

4. Insert a small straw into the hole you have just cut.

5. Attach the straw to a small hose, and attach the hose to an air outlet.

6. Turn on the air slowly and watch the lungs inflate. Crimp the hose to watch the lungs deflate.

✔ Check Your Understanding

1. In most cases of choking, the obstruction is caught in the right primary bronchus. Considering the shape and size of each primary bronchus, explain why the obstruction tends to lodge in the right rather than the left primary bronchus.

2. The right primary bronchus, as noted in the previous question, is the most common site for an obstruction. When we perform a cricothyroidotomy, are we bypassing this blockage? What are we bypassing to restore ventilation?

Part 2: Inflating Fresh Lungs

In the second part of this exercise, you will examine lung inflation and deflation in a fresh specimen rather than a preserved specimen.

Procedure

1. Obtain a fresh specimen and a large air hose.

2. Examine the specimen for structures covered in Exercise 1 —in particular the structures of the larynx, including the epiglottis and vocal folds, the trachea and its hyaline cartilage rings, and the pleura.

3. Note the texture of the lungs, and compare it with that of the lungs of the fetal pig:

4. Insert the air hose into the larynx and feed it down into the trachea. Take care not to get the hose stuck in one of the primary bronchi.

5. Attach the hose to the air outlet and turn it on slowly. You may have to squeeze the trachea and the hose to prevent air from leaking out.

6. Observe as the lungs inflate. You may inflate the lungs quite full without running the risk of having the lungs rupture.

7. Note the texture of the full lungs:

8. Crimp the air hose and watch the lungs deflate. Again, feel the lungs and note changes in texture:

22. Respiratory System Physiology

■ **OBJECTIVES**

Once you have completed this unit, you should be able to:

1. Describe the pressure-volume relationships in the lungs that permit inspiration and expiration.

2. Describe and measure respiratory volumes and capacities.

3. Measure the difference in the carbon dioxide content of exhaled air before and after exercising, and relate this to effects on pH.

■ **MATERIALS**

- Bell jar model of lung
- Spirometer: bell ("wet") and/or computerized
- Disposable mouthpieces
- pH meter
- Drinking straws
- 200 mL glass beakers with deionized water

PRE-LAB EXERCISES

Prior to coming to lab, complete the following exercises, using your textbook for reference.

1 PRE-LAB EXERCISE 1:
Respiratory Volumes and Capacities

Complete the chart below by defining and giving the normal value for each of the following respiratory volumes and capacities.

VOLUME / CAPACITY	DEFINITION	NORMAL VALUE (IN mL)
Tidal volume		
Inspiratory reserve volume		
Expiratory reserve volume		
Residual volume		
Inspiratory capacity		
Functional residual capacity		
Vital capacity		
Total lung capacity		

2 PRE-LAB EXERCISE 2:
Respiratory Volumes and Capacities, *continued*

Color-code Figure 22.1, showing respiratory volumes and capacities.

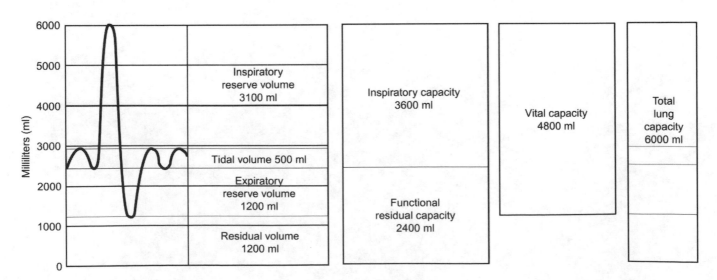

FIGURE 22.1
Diagram pulmonary volumes and capacities.

EXERCISES

In the previous unit, you learned about the structures of the respiratory system. Now, you will apply your knowledge of respiratory anatomy to respiratory physiology. As you go through this unit, remember that form always follows function, and that the anatomy of the respiratory system directly relates to its physiology.

The exercises in this unit will allow you to view the effects of pressure and volume on the lungs and to use a spirometer to measure respiratory volumes and capacities. In the final exercise, you will relate the pH of the blood to respiratory rate.

EXERCISE 1:
Pressure–Volume Relationships in the Lungs __

Ventilation, the physical movement of air in and out of the lungs, consists of two phases:

1. inspiration, and

2. expiration.

During *inspiration*, the volume of the lungs increases as air is brought into the lungs. During expiration, the volume decreases as air moves out of the lungs. The changes in volume that occur during inspiration and expiration are driven by changes in pressure. The relationship of gas pressure and volume is expressed in what is known as *Boyle's law*:

$$P_1V_1 = P_2V_2$$

Stated simply, this means that pressure and volume are inversely proportional: As the volume increases, the pressure decreases, and as the volume decreases, the pressure increases. With respect to ventilation, the inspiratory muscles (the diaphragm and the external intercostals) contract, which increases the height and diameter of the thoracic cavity.

Recall that the lungs are attached to the thoracic cavity directly by the pleural membranes. Therefore, as the thoracic cavity increases in size, the lungs increase in volume. As the volume increases, the intrapulmonary pressure decreases. When the pressure in the lungs is lower than the pressure in the atmosphere, air rushes into the lungs. Because the intrapulmonary pressure must be *lower* than the atmospheric pressure for inspiration to occur, it sometimes is referred to as *negative pressure*.

Expiration is achieved primarily by the elastic recoil of the lungs: Once the inspiratory muscles begin to relax, the elastic tissue of the lungs recoils, decreasing the volume of the lungs and increasing the pressure. Once the intrapulmonary pressure is higher than atmospheric pressure, air begins to exit the lungs. In the event of forced expiration,

several muscles, including the internal intercostals, will decrease the height and diameter of the thoracic cavity.

To view the effects of pressure and volume on the lungs, we will use a bell jar model of the lungs. The bell jar model has two balloons, each representing one lung, with a flexible membrane on the bottom, which represents the diaphragm.

Procedure

1. Apply upward pressure to the diaphragm. This represents how the diaphragm looks when it is relaxed. What has happened to the pressure of the system (has it increased or decreased)? What happened to the volume of the lungs?

2. Now slowly release the diaphragm. This represents the diaphragm flattening out as it contracts. What is happening to the pressure as you release the diaphragm? What happened to the volume of the lungs?

3. If your bell jar model has a rubber stopper in the top, you can use it to demonstrate the effects of a *pneumothorax* on lung tissue. A pneumothorax generally is caused by a tear in the pleural membranes that allows air to enter the pleural cavity. With the diaphragm flat and the lungs (balloons) inflated, loosen the rubber stopper. What happens to the lungs? Why?

✔ Check Your Understanding

1. Why would a pleural effusion, the presence of excess fluid in the pleural space, make inspiration difficult? (*Hint*: Think of the change in the intrapulmonary pressure that may result from a pleural effusion.)

2. The condition *emphysema* results in loss of elastic recoil of the lung tissue. Would this make inspiration or expiration difficult? Explain.

3. Why might you have difficulty breathing at higher altitudes, where the atmospheric pressure is low?

▋2 EXERCISE 2:
Measuring Respiratory Volumes and Capacities

Respiratory volumes, measured with an instrument called a *spirometer*, are useful tools with which to assess pulmonary function. They are especially helpful in differentiating the two primary types of respiratory disorders—restrictive diseases and obstructive diseases.

1. *Restrictive diseases*, such as pulmonary fibrosis, are characterized by a loss of elasticity (decreased compliance) of the lung tissue. As a result, patients' ability to inspire is affected adversely.

2. *Obstructive diseases*, such as chronic obstructive pulmonary disease (COPD) and asthma, are characterized by increased airway resistance, caused by narrowing of the bronchioles, increased mucus secretion, and/or an obstructing body such as a tumor. It may seem counterintuitive, but obstructive diseases make expiration, rather than inspiration, difficult. This is because, as expiration takes place, the increased intrapulmonary pressure naturally tends to shrink the diameter of the bronchioles. When the bronchioles are already narrowed, as in an obstructive disease, the increased intrapulmonary pressure can actually collapse the bronchioles, trapping oxygen-deficient air in the distal respiratory passages. Therefore, patients with obstructive diseases often exhale slowly and through pursed lips, to minimize the pressure changes and maximize the amount of air exhaled.

The normal values for respiratory volumes vary based upon age, sex, size, and physical condition.

Two types of spirometers are commonly found in anatomy and physiology labs: (1) a bell, or "wet" spirometer, and (2) a computerized spirometer. Many types of wet spirometers allow you to assess only expiratory volumes, whereas computerized spirometers typically allow you to assess both inspiratory and expiratory volumes. Both kinds of apparatus allow you to measure one or more respiratory *capacities*, which are obtained by adding together two or more respiratory volumes. Capacities are useful because they give a more complete picture of pulmonary function. Refer to the Pre-Lab Exercises to review the definitions and average values for each respiratory volume and capacity.

The procedure below is intended for a wet spirometer, which allows measurement of expiratory volumes only. If your lab has a computerized spirometer, you can either follow the procedure below or follow the prompts by the computer program.

Procedure: "Wet" Spirometer

> ➤ NOTE: IF YOU HAVE CARDIOVASCULAR OR PULMONARY DISEASE, OR ARE PRONE TO DIZZINESS OR FAINTING, YOU ARE CAUTIONED AGAINST ENGAGING IN THE FOLLOWING ACTIVITY.

1. Obtain a disposable mouthpiece and attach it to the end of the tube.

2. Before you begin, practice exhaling through the tube several times. Note that you are supposed to only *exhale* into the tube, as the tube likely has no filter.

3. *Measure the tidal volume*: Take in a normal breath, and exhale this breath through the tube. Getting a true representation of the tidal volume is often difficult, because people have a tendency to force the expiration. To get the most accurate tidal volume, take several measurements and average the numbers.

Measurement 1: _____

Measurement 2: _____

Measurement 3: _____

Average Tidal Volume: _____

4. *Measure the expiratory reserve volume*: Before taking this measurement, inhale and exhale a series of tidal volumes. Then take in a normal tidal volume, breathe out a normal tidal volume, put the mouthpiece to your mouth, and exhale as forcibly as possible. Don't cheat and take in a large breath first! As before, perform several measurements and average the numbers.

Measurement 1: _____

Measurement 2: _____

Measurement 3: _____

Average Expiratory Reserve Volume: _____

5. *Measure the vital capacity*: As before, inhale and exhale a series of tidal volumes. Then bend over and exhale maximally. Once you have exhaled as much air as you can, raise yourself upright and inhale as much air as you possibly can (or, as I tell my students, until you feel like you are about to "pop"). Quickly place the mouthpiece to your mouth and exhale as forcibly and as long as possible. Take several measurements (you may want to give yourself a minute to rest between measurements), and record the data below.

Measurement 1: _____

Measurement 2: _____

Measurement 3: _____

Average Vital Capacity: _____

6. *Calculate the inspiratory reserve volume*: Even though the spirometer cannot measure inspiratory volumes, you can calculate this volume now that you have the vital capacity (VC), tidal volume (TV), and expiratory reserve volume (ERV). Recall that VC = TV + IRV + ERV. Rearrange the equation: IRV = VC − (TV + ERV).

Average IRV: _____

7. How do your values compare with your classmates' values? What factors, if any, do you think may have affected your results?

✔ **Check Your Understanding**

1. Of the pulmonary volumes and capacities, which do you think would be most affected by a restrictive disease? Why?

2. Which of the pulmonary volumes and capacities do you think would be most affected by an obstructive disease? Why?

**3 EXERCISE 3:
pH and Ventilation** _____

Recall that ventilation is required not only to bring oxygen into the system but also to *release carbon dioxide*. Carbon dioxide, which is a nonpolar gas that does not dissolve well in plasma, is transported through the blood mainly as *bicarbonate*. The process in which carbon dioxide reacts with water to produce carbonic acid, which subsequently dissociates into bicarbonate and a hydrogen ion, is catalyzed by an enzyme called carbonic anhydrase. The reaction, the reverse of which occurs in the lungs to reform carbon dioxide and water, occurs as:

$$H_2O + CO_2 \rightleftharpoons H_2CO_3 \rightleftharpoons HCO_3^- + H^+$$

The continual formation of carbonic acid and bicarbonate is vital, as it constitutes one of the primary *buffer* systems in the body. Recall that a buffer is a chemical that resists changes in pH. This is extremely important for maintaining homeostasis, as the pH of the blood must stay in the range of 7.35–7.45 for most enzymes and proteins to function.

It follows that, because carbon dioxide is an important factor in maintaining the pH of the blood and carbon dioxide is eliminated from the body by the lungs, the rate and depth of ventilation are critical in determining the level of carbon dioxide, and therefore the pH, of the blood. The following simple experiment will allow you to see the effects of carbon dioxide on pH and the effects of carbon dioxide on ventilatory rate.

Procedure

1. Measure your partner's respiratory rate while he/she is seated and resting. Measuring the respiratory rate can be tricky because as soon as you tell your partner to breathe normally, he/she automatically becomes conscious of breathing and alters the rate. A trick I learned is to place two fingers on the radial artery and take the pulse. While you are doing this, your partner is more focused on the pulse rate, and he/she breathes more naturally. Record this value in the chart below.

2. Obtain a 200mL beaker about half-full of deionized water. Measure the pH of the water with a pH meter. Record this value below:

    pH of deionized water: _____

3. Place a drinking straw into the water and have your partner exhale through the straw into the water (the water should bubble) for 20 seconds. Then measure the pH again.

4. Have your partner do vigorous exercise for at least 5 minutes (running up and down stairs usually does the job quite nicely). Immediately upon finishing the exercise, have your partner blow through the straw into the beaker of water again for 20 seconds. Measure the pH with a pH meter.

5. After you record the pH, measure the respiratory rate (you may want to use the pulse trick again).

6. Allow your partner to have a (well-deserved) rest for 5 more minutes. Again, have your partner blow through the straw into the beaker of water for 20 seconds, and measure the pH. Also measure the respiratory rate. Record all of your data in the chart below.

SETTING	pH	RESPIRATORY RATE
At rest		
Immediately after exercise		
After 5 minutes of rest		

7. Let's figure out what happened, and why:
    a. Start with the basics: What happens to the metabolism during exercise? What is generated as a byproduct of glucose metabolism?

b. What does this byproduct do to the pH of the blood? Why?

c. What sort of ventilatory response does this change in pH cause (hyperventilation or hypoventilation)? Why?

d. Now apply this to your results:
    (1) What happened to the respiratory rate during and after exercise? Why? (*Hint*: See item c. above.)

    (2) What happened to the pH of the water after exercise? Why?

    (3) What happened to the respiratory rate and the pH after resting? Why?

✔ Check Your Understanding

1. When a person is hyperventilating, why is it recommended that he/she breathe into a paper bag?

2. A patient presents in a state of *ketoacidosis* (a type of metabolic acidosis in which the pH of the blood drops), after going on a low-carb, all-fat diet. Will the patient be hyperventilating, or hypoventilating? Why?

23. Urinary System Anatomy

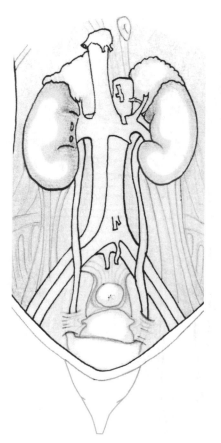

■ OBJECTIVES

Once you have completed this unit, you should be able to:

1. Identify anatomical structures of the urinary system.

2. Perform a dissection of a fresh or preserved kidney.

3. Identify microscopic structures of the urinary system on prepared histological slides.

■ MATERIALS

- Anatomical models: kidney, nephron
- Microscope slides: transitional epithelium, kidney cortex, kidney medulla
- Fresh or preserved kidneys
- Dissection equipment
- Colored pencils

PRE-LAB EXERCISES

Prior to coming to lab, use your text and lab atlas to complete the following exercises.

1 PRE-LAB EXERCISE 1: Kidney Anatomy

Figure 23.1 shows two diagrams of the kidney. Label and color-code the diagrams, using the terms from Exercise 1 (please note that all structures may not be visible in the diagrams). See Figures 16.1 and 16.2 in your lab atlas for reference.

2 PRE-LAB EXERCISE 2: Nephron

Figure 23.2 is a diagram of the nephron. Label and color-code the diagram, using the terms from Exercise 1 (please note that all structures may not be visible in the diagram). See Figure 16.3 in your lab atlas for reference.

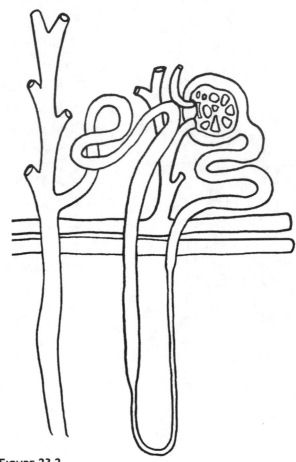

FIGURE 23.2
Diagram of nephron.

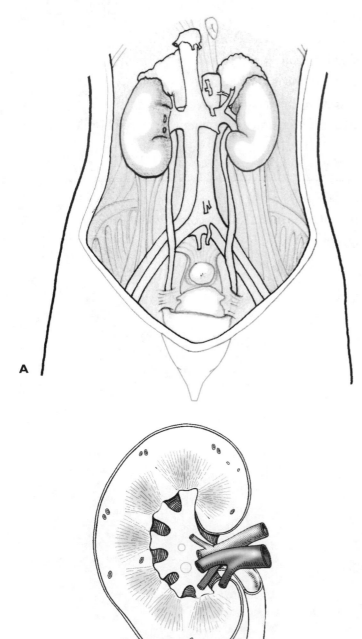

FIGURE 23.1
(A) Diagram of the urinary system, and (B) frontal section of a kidney.

EXERCISES

The urinary system consists of the paired kidneys, ureters, and single urinary bladder and urethra. The primary functions of the kidneys include removing nitrogenous wastes from the body and maintaining fluid, electrolyte, and acid/base balance. The kidneys also produce the hormone erythropoietin, which regulates hematopoiesis, help the liver detoxify certain compounds, and make glucose during times of starvation. The ureters, urinary bladder, and urethra transport and temporarily store urine until it is eliminated from the body during a process known as micturition.

The exercises in this unit will help you get acquainted with the anatomy of the urinary system on models and preserved specimens. You will also examine the histology of the kidney and urinary bladder, and relate it to the function of the urinary system.

EXERCISE 1:
Identification of Urinary Structures

The kidneys are situated against the posterior body wall, in a *retroperitoneal* position (so named because they are behind the peritoneal membranes that surround many of the digestive organs). They are encased within three layers of connective tissue, the thickest of which is a middle layer of adipose tissue. The adipose tissue helps to wedge the kidneys in place; a potential consequence of starvation is the resulting loss of this adipose tissue, causing the kidneys to droop, kinking the ureter (a condition known as *nephroptosis*).

Internally, the kidney is divided into three regions:

1. *Renal cortex*: outer region, which consists of many blood vessels that serve the functional units of the kidney, the *nephrons*. Most of the nephron is located in the cortex.

2. *Renal medulla*: middle region, consisting of triangular *medullary pyramids* separated by structures called *renal columns* (these resemble the cortex in appearance). The pyramids have a striated (striped) appearance, because of the preponderance of small tubules, the loops of Henle and the collecting ducts.

3. *Renal pelvis*: this inner region, which houses a basin for collecting urine drained from the pyramids. The urine drains from the renal pelvis into the *ureter*.

The blood supply of the kidney is unique. Macroscopically, it begins with the large *renal arteries*, which carry about 1200 milliliters per minute to the kidney to be filtered. The renal arteries branch into progressively smaller arteries as they pass through the medulla to the cortex. In the cortex, the vessels branch into small *afferent arterioles*. The afferent arterioles supply a tuft of capillaries known as the *glomerulus*, where the blood is filtered.

We have seen that normally a capillary bed drains into a venule; however, this is not the case with the microcirculation of the kidney. Instead of draining into a venule, the glomerulus drains into a second arteriole called the *efferent arteriole*. The efferent arteriole then gives rise to a second capillary bed, the *peritubular capillaries*. These capillaries surround the tubules of the nephron, and pick up substances that have been reabsorbed from the tubules. This second capillary bed then drains out through the venous system of the kidney.

Microscopically, the kidney is composed of more than a million tiny units called *nephrons*. The first part of the nephron, where blood is filtered, is called the renal corpuscle, and it consists of the glomerulus and the surrounding glomerular (or Bowman's) capsule. Filtered fluid, called filtrate, first enters a space known as the capsular space, which is found within a structure called the *glomerular capsule* (also known as *Bowman's capsule*). The glomerular capsule consists of two layers: (a) the parietal layer, which is simple squamous epithelium, and (b) the visceral layer, which consists of cells called *podocytes* that surround the capillaries of the glomerulus. The podocytes have extensions called foot processes that interlock to form thin *filtration slits*. The filtration slits prevent large substances in the blood, such as blood cells and proteins, from entering the filtrate.

From the glomerular capsule the filtrate enters the *renal tubule*, which consists of three parts: (1) the *proximal convoluted tubule*, (2) the *nephron loop* (also called the *loop of Henle*), and (3) the *distal convoluted tubule*. Several distal convoluted tubules drain into one *collecting duct*.

As the filtrate flows through the renal tubule and collecting duct, it is modified and most of the water and solutes are reclaimed. After the filtrate drains from the collecting ducts into larger tubules called *papillary ducts*, which finally drain into larger *minor calyces*, it is called *urine*. The minor calyces drain into *major calyces*, which empty into the renal pelvis. From the pelvis urine enters muscular tubes called the *ureters*, which massage the urine via peristalsis down to the posteroinferior wall of the *urinary bladder*. The inferior portion of the urinary bladder, the *trigone*, contains the internal urethral orifice, which continues into the *urethra*, where the urine exits the body.

In this laboratory period you will identify the structures outlined below.

Kidney Structures
(see Figures 16.1, 16.2, and 16.4–16.7 in your lab atlas)

1. Surrounding connective tissue:
   a. Renal fascia
   b. Adipose capsule
   c. Renal capsule
2. Hilum
3. Regions:
   a. Renal cortex
   b. Renal medulla
      (1) Renal columns
      (2) Renal pyramids
      (3) Renal papilla
   c. Minor calyces
   d. Major calyces
   e. Renal pelvis
4. Blood supply:
   a. Renal artery
      (1) Segmental artery
      (2) Lobar artery
      (3) Interlobar artery
      (4) Arcuate artery
      (5) Interlobular artery
   b. Renal vein
      (1) Interlobular vein
      (2) Arcuate vein
      (3) Interlobar vein

Microanatomy: Nephron
(see Figure 16.3 in your lab atlas)

1. Renal corpuscle
   a. Glomerulus
      (1) Afferent arteriole
      (2) Efferent arteriole
      (3) Peritubular capillaries
      (4) Vasa recta
   b. Glomerular (Bowman's) capsule
      (1) Parietal layer
      (2) Visceral layer
      (3) Podocytes
2. Renal tubule
   a. Proximal convoluted tubule
   b. Nephron loop (loop of Henle)
      (1) Descending limb
      (2) Ascending limb
   c. Distal convoluted tubule
3. Juxtaglomerular apparatus
   a. JG cells
   b. Macula densa
4. Collecting duct
5. Papillary duct

Other Structures of the Urinary System
(see Figures 16.1, 16.2, and 16.4–16.7 in your lab atlas)

1. Ureter
2. Urinary bladder
   a. Detrusor muscle
   b. Ureteral orifices
   c. Internal urethral orifice
   d. Trigone
3. Urethra
   a. Internal urethral sphincter
   b. External urethral sphincter
   c. External urethral orifice

Model Inventory

As you view the anatomical models and torsos in lab, list them on the inventory below, and state which of the previous structures you are able to locate on each model.

MODEL/DIAGRAM	STRUCTURES IDENTIFIED

✔ Check Your Understanding

1. Recall from our discussion of the lymphatic system that the systemic capillary beds, which are fed by an arteriole and drained by a *venule*, lose about 1.5 mL/minute of fluid into the interstitial space. By contrast, the glomerular capillaries, which are both fed and drained by arterioles, lose about 120 mL/minute into the capsular space. This value is called the *glomerular filtration rate* (GFR). What do you think would happen to the GFR if:

   a. The afferent arteriole underwent vasodilation?

   b. The afferent arteriole underwent vasoconstriction?

2. The epithelial tissue of the proximal convoluted tubule is simple cuboidal epithelium with microvilli. Why do you think microvilli are present in this section of the tubule?

6. Use a scalpel to make a frontal section of the kidney. List the internal structures you are able to identify:

2 EXERCISE 2:
Kidney Dissection _____

In this exercise we will identify several of the gross structures of the kidney from the previous exercise.

Procedure

> NOTE: SAFETY GLASSES AND GLOVES ARE REQUIRED.

1. Obtain a fresh or preserved kidney specimen and dissection supplies.

2. If the thick surrounding connective tissue coverings are intact, note their thickness and amount of adipose tissue.

3. Use scissors to cut through the capsule and remove the kidney.

4. List surface structures that you are able to identify:

5. Distinguishing between the ureter, the renal artery, and the renal vein is often difficult. Following are some hints to aid you in this process.
   a. The renal artery typically has the thickest and most muscular wall, and prior to entering the kidney, it branches into segmental arteries.
   b. The renal vein is thinner and flimsier, and often larger in diameter than the renal artery. Once it has left the kidney, it does not branch.
   c. The ureter has a thick, muscular wall, too, but it does not branch once it leaves the kidney. Also, its diameter is usually smaller than either the renal artery or the renal vein.

Keeping these points in mind, determine the location of the renal artery, the renal vein, and the ureter on your specimen.

3 EXERCISE 3:
Urinary Histology _____

In this exercise we will examine three sections of urinary tissue: (1) *transitional epithelium*, from the urinary bladder, (2) the renal cortex, and (3) the renal medulla. When examining these tissue sections, pay attention to the following:

● Transitional epithelium is often stained with a blue dye, which makes it easy to identify on a lab practical. If it isn't stained blue, it is still easy to identify because of the unique shape of the cells. Recall that this is stratified epithelium, so you will see apical and basal cells. Notice that the cells at the apical edge are dome-shaped (or sometimes squamous in appearance) and that those nearer the basal edge are typically cuboidal. When examining the transitional epithelium, start on low power, identify the edge with the epithelial tissue, then move to medium and/or high power. Once you have examined the epithelial tissue, scroll down the slide to see layers of connective tissue and smooth muscle.

● In the slide of the renal cortex, you will notice the preponderance of small, ball-shaped structures. These are glomeruli. They are surrounded by a space lined by a ring of simple squamous epithelial cells. These cells are the parietal layer of the glomerular capsule, and the space is the capsular space, where the filtrate first enters the nephron. In the area around the glomeruli and the glomerular capsule, you will see cross-sections of nephron tubules (the proximal and distal convoluted tubules and sections of the loops of Henle).

● Recall that the renal medulla is composed mostly of collecting ducts and the loops of Henle. You will not see any glomeruli in the medulla, as they are confined to the cortex.

Examine the following tissue sections and, using colored pencils, draw what you see and label your drawing with the structures indicated.

Transitional epithelium (urinary bladder)
(See Figure 16.15 in your lab atlas.)

Label the following on your drawing:
1. Transitional epithelium
2. Muscularis (detrusor muscle)

Kidney (renal) Cortex
(See Figures 16.12 and 16.13 in your lab atlas.)

Label the following on your drawing:
1. Glomerulus
2. Glomerular capsule with simple squamous epithelium
3. Capsular space
4. Kidney tubules

Kidney (renal) Medulla
(See Figure 16.10 in your lab atlas.)

Label the following on your drawing:
1. Loop of Henle (in cross-section)
2. Collecting duct

✔ Check Your Understanding

1. Why do you think the epithelium of the urinary bladder is stratified? Why do you think that the cells are dome-shaped rather than squamous, columnar, or cuboidal? (*Hint*: Remember that form follows function.)

2. One of the structures you labeled on your drawing is the capsular space between the parietal and visceral layers of the glomerular capsule. Under normal conditions, would you expect to see red or white blood cells in this space? Why or why not?

24. Urinary System Physiology

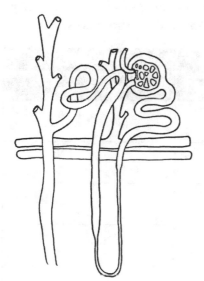

■ **OBJECTIVES**

Once you have completed this unit, you should be able to:

1. Model the physiology of the kidney, and test for chemicals that are in the filtrate.

2. Perform and interpret urinalysis on simulated urine specimens.

3. Research the possible causes of abnormalities detected on urinalysis.

4. Trace an erythrocyte, a glucose molecule, and a urea molecule through the gross and microscopic anatomy of the kidney.

■ **MATERIALS**

- Dialysis tubing
- Animal blood or simulated blood
- Deionized water
- Urinary Reagent Strips-10 (URS-10) vials
- 100 mL beaker
- Graduated cylinder
- String
- Simulated urine samples
- pH paper

PRE-LAB EXERCISES

Prior to coming to lab, use your text and lab atlas to complete the following exercises.

1 | PRE-LAB EXERCISE 1: Structure of the Nephron_____

Review the structure of the nephron by labeling and color-coding the diagram in Figure 24.1.

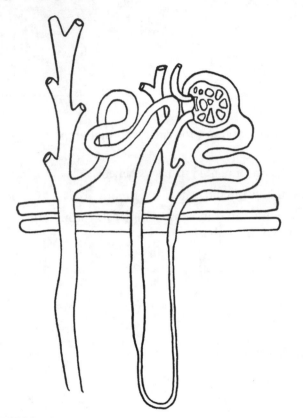

FIGURE 24.1
Diagram of a nephron.

2 | PRE-LAB EXERCISE 2: Glomerular Filtration and Reabsorption_____

Below is a chart listing chemicals and cells found in blood. Use your text to determine whether the substance is filtered at the glomerulus. If it is filtered, determine if the substance should be found in the urine, or if it is reabsorbed into the blood in the nephron. If it is reabsorbed, further determine where in the nephron reabsorption of this substance occurs. The chart should prove to be useful in this unit, as we will be testing urine samples for many of these substances.

SUBSTANCE	FILTERED? (YES/NO)	REABSORBED (YES/NO) IF YES, WHERE IN THE NEPHRON IS IT REABSORBED?
Erythrocyte		
Leukocyte		
Water		
Glucose		
Proteins		
Amino acids		
Urea		
Creatinine		
Electrolytes (sodium, potassium, chloride)		
Uric acid		
Ketones		

EXERCISES

In the last unit you learned that one major function of the nephron is filtration; substances are filtered at the glomerulus, and filtrate is formed at a rate of 120 ml/min. A second major function of the glomerulus is to reclaim much of what is filtered, including 99% of the water in the filtrate and most of the solutes. The importance of this function cannot be understated. If the fluid were not reabsorbed in the renal tubule and the collecting duct, we would lose our entire plasma volume in under 30 minutes!

To examine the processes of filtration, you will model a kidney and test for substances in the "filtrate" in the first exercise. In the second exercise you will conduct a commonly performed medical test, a urinalysis. You will test several urine samples and determine what potential pathological conditions could cause the detected abnormalities. In the final exercise you will trace the pathway of molecules through the general circulation and the microanatomy of the kidney in order to contrast the filtration and reabsorption of three substances: red blood cells, urea, and glucose.

EXERCISE 1: Modeling a Kidney

In the Pre-Lab Exercises you determined which substances enter the filtrate and which are not filtered and, therefore, stay in the glomerulus. In this exercise we will demonstrate this firsthand by modeling a kidney and testing for substances that appear in the filtrate. We will construct a model kidney using simple dialysis tubing, which has a permeability similar to that of the glomerulus. We will place in the dialysis tubing either animal blood or simulated blood that contains all of the normal components of blood, including erythrocytes, leukocytes, proteins, and glucose. We will then immerse the blood-filled tubing in water and see which substances are permitted to leave the tubing to enter the surrounding water.

To interpret our results, we will use a urinalysis reagent test strip (URS-10) (Figure 24.2). These strips allow us to analyze the components of urine. They consist of 10 small, colored pads that change color in the presence of certain chemicals. To interpret the strip after it has been immersed in urine, you will be watching the pads for color changes and comparing the color changes to a color-coded key on the side of the bottle. The color that is closest to that on the strip is recorded as your result. Please note that for this exercise we will not read all 10 boxes but, instead, only four: blood, leukocytes, protein, and glucose.

Procedure

> ➤ NOTE: SAFETY GLASSES AND GLOVES ARE REQUIRED.

1. Cut a 4-inch piece of dialysis tubing.
2. With string, securely tie off one end of the tubing, and open the other end of the tubing by wetting the end of the tube and rubbing it between your fingers.
3. Fill the dialysis tubing about half-full with either animal blood or simulated blood, and securely tie off the open end of the tube with string.
4. Place the tied-off tube in a beaker containing about 50 mL of deionized water.
5. Leave the tubing in the water for approximately 25 minutes. While you are waiting, move on to the next exercise.
6. After 25 minutes, remove the tubing from the water.
7. Use a URS-10 to test for the presence of the following substances in the water, which models the "filtrate." Dip the strip in the filtrate and immediately remove it. Turn the bottle on its side and compare the colors of the pads for glucose, blood, leukocytes, and protein. You will notice that on the side of the bottle, time frames are listed for each substance that is tested. It is best to wait this amount of time to watch for a reaction; otherwise, you could obtain a false negative result. If you wait too long to read the results, though, the colors will tend to darken and may blend with adjacent colors.
8. Record your results below.

    a. Glucose:

    b. Erythrocytes (blood):

    c. Leukocytes:

    d. Protein:

FIGURE 24.2
URS-10 vial and strip.
Reprinted with permission by Craig Medical Distribution.

9. Which substances were filtered out? Which substances stayed in the tubing?

✔ Check Your Understanding

1. Why were certain substances filtered out while others remained in the simulated glomerulus?

2. Of the substances for which you tested, should any of these appear in the urine? Why or why not?

2 EXERCISE 2: Urinalysis

For hundreds of years, health care providers have recognized the utility of urinalysis as a diagnostic tool. Historically, the urine was evaluated for color, translucency, odor, and taste (yes, taste!). Today, while these characteristics are still examined (except, thankfully, taste), we also utilize urinalysis test strips, as we did in Exercise 1, to test for the presence of various chemicals in the urine. Abnormal readings, while not necessarily diagnostic of a specific condition, can give the provider a wealth of information about the state of a patient's renal function and overall health.

In this exercise we will analyze different urine samples using a URS-10, as in Exercise 1. For each urine sample, one or more results will be read as abnormal. In the second part of the exercise, you will research potential causes of the abnormalities that you detected.

Procedure: Part One

> ➤ NOTE: SAFETY GLASSES AND GLOVES ARE REQUIRED.

1. Randomly choose one sample of simulated urine and pour approximately 3 mL into a test tube or graduated cylinder.

2. Obtain a piece of pH paper and put a small drop of the simulated urine on the paper. Compare the color on the pH paper with the colors on the container of pH paper to determine the pH. Record your result in the chart. Please note that the URS-10 also detects pH; however, a more accurate reading is obtained by using pH paper.

3. Submerge one URS-10 in the urine, and immediately remove it.

READING	SAMPLE 1	SAMPLE 2	SAMPLE 3	SAMPLE 4	SAMPLE 5
pH					
Leukocytes					
Nitrite					
Urobilinogen					
Protein					
Blood					
Specific gravity					
Ketones					
Bilirubin					
Glucose					

4. Compare the resulting colors and patterns on the test strip to those on the key on the bottle, taking care to wait the appropriate amount of time to read the results.

5. Record your results in the chart, noting any anomalous results.

6. Repeat for the remaining samples.

Procedure: Part Two

Now take your results and find out what could potentially lead to those specific abnormalities on a urinalysis. Although your textbook likely will have some information about this question, you will find more complete information by going to the Internet.

SAMPLE	PRIMARY ABNORMALITY	POTENTIAL CAUSES
1		
2		
3		
4		
5		

✔ Check Your Understanding

1. You probably noticed in your research that findings of blood and protein in the urine can be indicative of renal failure stemming from inflammation of the glomerulus. Using your knowledge of the inflammatory response, explain why substances that normally wouldn't enter the filtrate are found in the urine when the glomerulus is inflamed.

2. A common reason for finding glucose in the urine is poorly controlled diabetes mellitus. Explain this finding.

3 EXERCISE 3: Tracing Substances Through the Kidney

It's time to trace again! Here we will trace the pathway of different molecules through the general circulation and then through the vasculature and microanatomy of the kidney. We will examine three different substances:

1. a red blood cell,

2. a molecule of glucose, and

3. a molecule of urea.

Following are some hints to use as you trace:

● Refer to Unit 17 (Blood Vessel Anatomy) to review the pathway of blood flow through the body.

● Don't forget the basic rules of blood flow: If you start in a capillary bed (which you will for each of these), you first must go through a *vein*, and then the heart and pulmonary circuit, before you may enter the arterial system.

● As you trace each substance, refer to the chart in Pre-Lab Exercise 2. This will help you to determine if the substance gets filtered, if a filtered substance gets reabsorbed, and where the reabsorption occurs.

● Refer to your textbook for a diagram of the microcirculation and microanatomy of the kidney.

Part One: Erythrocyte

Trace an erythrocyte from its origin in the bone marrow of the left humeral head to the renal *vein*. Keep in mind that the erythrocyte will first enter a capillary bed in the bone marrow before it is drained by a local vein.

Part Two: Glucose

Trace a molecule of glucose from its point of absorption into the capillaries of the small intestine to the renal vein. Don't forget the hepatic portal system!

Part Three: Urea

Trace a molecule of urea from its origin in the liver to its final destination outside the body.

25. Digestive System

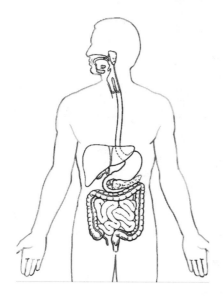

■ **OBJECTIVES**

Once you have completed this unit, you should be able to:

1. Identify structures of the digestive system on models, diagrams, and preserved small mammals.

2. Identify histological structures of the digestive system on microscope slides.

3. Trace the pathway of physical and chemical digestion of carbohydrates, proteins, and lipids.

4. Demonstrate the action of emulsifying agents on lipids.

■ **MATERIALS**

● Anatomical models: digestive system, head and neck, human torso, human skulls with teeth

● Microscope slides: tooth, esophagus, gastroesophageal junction, stomach—fundus, duodenum, ileum with Peyer's patches, pancreas, liver

● Fetal pigs with dissecting trays and kits

● Colored pencils

● Laminated outline of human body and water-soluble marking pens

● Detergent

● Distilled water

● Vegetable oil

● Red Sudan stain

PRE-LAB EXERCISES

Prior to coming to lab, use your text and lab atlas to complete the following exercises.

1 PRE-LAB EXERCISE 1:
Anatomy of the Digestive System _____

Figure 25.1 illustrates the organs of the digestive system. Use colored pencils to color-code the organs and structures of the digestive system (you may choose to wait until the laboratory period to color-code the structures in a manner similar to the digestive models you use in lab). Next, label each of the organs you have color-coded using the terms from Exercise 1. (Please note that not all structures in the list from Exercise 1 will be visible in this diagram.) For reference, see Figures 15.1 and 15.16 in your lab atlas.

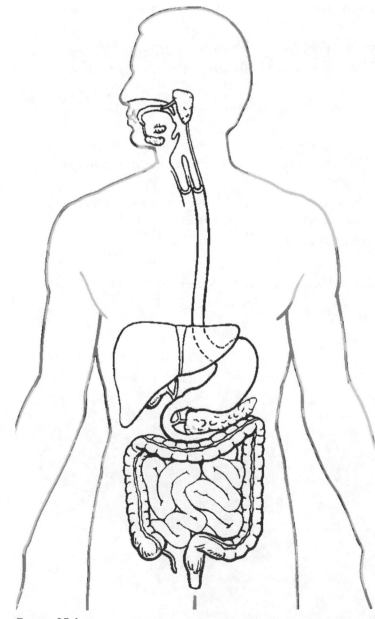

FIGURE 25.1
Organs of the digestive system.

2 PRE-LAB EXERCISE 2:
Digestive Enzymes _____

The following chart lists digestive enzymes. Consult your text to determine the organ that produces the enzyme, along with the function of the enzyme. You will use this information in lab.

ENZYME	SOURCE	FUNCTION
Salivary amylase		
Pepsin		
Trypsin		
Chymo-trypsin		
Carboxy-peptidase		
Amino-peptidase		
Pancreatic lipase		
Pancreatic amylase		
Maltase		
Lactase		
Sucrase		
Nuclease		

3 PRE-LAB EXERCISE 3:
Lipid Emulsification and Digestion _____

During this laboratory period we will examine the process of emulsification of lipids. Use your text and lab atlas to complete the following questions pertaining to lipid digestion and emulsification.

1. Define emulsification.

2. Why is emulsification necessary in the digestion of lipids?

3. Describe the composition of bile. Are all of the components of bile necessary for emulsification?

4. Describe the structure of bile salts. Explain how this structure permits bile salts to emulsify fats.

EXERCISES

The food that we eat contains nutrients that our cells use to build and repair body tissues, and to make energy in the form of ATP. However, food macromolecules are typically too large for the body to absorb and utilize, so breaking them down into smaller molecules is required. This process of breaking down foods into smaller substances that can enter body cells is called _digestion_, and is carried out by the _digestive system_. In general, the functions of the digestive system include taking in food, breaking food down into nutrients, absorbing these nutrients into the bloodstream, and finally, eliminating indigestible substances.

The exercises in this unit will introduce you to the anatomy and histology of the digestive system. In addition, we will examine the physiological process of emulsification, and trace nutrients from their ingestion in the mouth to their absorption in the small intestine.

1 EXERCISE 1:
Digestive System Anatomy _____

The digestive system is composed of two types of organs:

1. _Organs of the alimentary canal_ (also known as the gastrointestinal, or GI, tract): organs through which food travels—the _mouth_, _pharynx_, _esophagus_, _stomach_, and _small_ and _large intestines_. In the alimentary canal, food is ingested, broken into smaller pieces both mechanically and chemically, moved from one section of the alimentary canal to the next (a process called _propulsion_), and the nutrients absorbed into the bloodstream.

2. _Accessory organs_: organs that assist in mechanical or chemical digestion—the _teeth_, _tongue_, _salivary glands_, _pancreas_, _liver_, and _gallbladder_. Most accessory digestive organs do not come into direct contact with the food (the teeth and tongue are an exception). Many secrete substances such as bile salts and enzymes that travel by way of a duct to the alimentary canal that assist in food breakdown.

Many of the digestive organs reside inside a cavity known as the _peritoneal cavity_. Like the pleural and pericardial cavities, the peritoneal cavity is found between a double-layered serous membrane consisting of the following layers:

1. _Parietal peritoneum_: a thin membrane that is functionally fused to the abdominal wall.

2. _Visceral peritoneum_: a thin membrane that adheres to the surface of the digestive system organs. Surrounding

some of the digestive organs—in particular, the intestines —the visceral peritoneum folds upon itself to form a thick membrane known as the *mesentery*. The mesentery houses blood vessels, nerves, and lymphatic vessels, and functions to anchor them in place.

Between the parietal and visceral peritoneal membranes is a layer of *serous fluid* that allows the organs to slide over one another without friction.

The following is a list of structures you will identify in this laboratory period:

Alimentary Canal

(See Figures 15.1, 15.16–15.22, 15.28–15.30, and 15.36–15.40 in your lab atlas.)

1. Mouth:
   a. Lips
   b. Cheeks
   c. Vestibule
   d. Oral cavity proper
   e. Hard palate
   f. Soft palate
   g. Uvula
2. Pharynx
   a. Oropharynx
   b. Laryngopharynx
3. Esophagus
4. Esophageal hiatus
5. Stomach
   a. Cardioesophageal sphincter (or lower esophageal sphincter)
   b. Cardia
   c. Fundus
   d. Body
   e. Antrum
   f. Pylorus
   g. Pyloric sphincter
   h. Rugae
6. Small intestine
   a. Duodenum
   b. Jejunum
   c. Ileum
   d. Ileocecal valve
   e. Plicae circulares
7. Villus
   a. Intestinal capillaries
   b. Lacteal
   c. Intestinal crypts
8. Large intestine
   a. Haustra
   b. Taeniae coli
   c. Epiploic appendages
   d. Cecum with vermiform appendix
   e. Ascending colon
   f. Hepatic flexure
   g. Transverse colon
   h. Splenic flexure
   i. Descending colon
   j. Sigmoid colon
   k. Rectum
   l. Anal canal

Accessory Organs

(See Figures 15.2–15.4, 15.43–15.45, 15.49, and 15.51 in your lab atlas.)

1. Teeth
2. Tongue
   a. Filiform papillae
   b. Fungiform papillae
   c. Circumvallate papillae
3. Salivary glands
   a. Parotid gland with parotid duct
   b. Submandibular gland
   c. Sublingual gland
4. Pancreas
   a. Main pancreatic duct
5. Liver
   a. Right lobe, left lobe, caudate lobe, and quadrate lobe
   b. Falciform ligament
   c. Hepatic duct
   d. Hepatic portal vein
   e. Hepatic arteries
   f. Hepatic veins
6. Gallbladder
   a. Cystic duct
   b. Common bile duct

Other Structures

1. Peritoneal cavity
2. Peritoneum
   a. Visceral peritoneum
   b. Parietal peritoneum
   c. Mesentery
3. Greater omentum
4. Blood vessels:
   a. Left gastric artery
   b. Gastric veins
   c. Superior mesenteric artery
   d. Inferior mesenteric artery

e. Superior mesenteric vein

f. Inferior mesenteric vein

> ➤ NOTE: YOUR INSTRUCTOR MAY WISH TO OMIT CERTAIN STRUCTURES INCLUDED ABOVE OR ADD STRUCTURES NOT INCLUDED IN THESE LISTS. LIST ANY ADDITIONAL STRUCTURES BELOW:

Model Inventory

As you view the anatomical models in lab, list them on the inventory below, and state which of the preceding structures you are able to locate on each model.

Model/Diagram	Structures Identified

✔ Check Your Understanding

1. Where does all nutrient-rich blood from the superior and inferior mesenteric veins, splenic vein, and gastric veins go before it enters the general circulation? Why?

2. One of the common consequences of gallstones is blockage of the common bile duct, which prevents bile from being emptied into the duodenum. This produces the easily recognizable symptom of "clay-colored" (non-pigmented) feces. Why do you think that this symptom is present, given the composition of bile? *(Hint: What is found in bile besides bile salts?)*

2 EXERCISE 2: Digestive System Histology _____

The organs of the alimentary canal are ideal for observing different tissue layers. The histology of the tissue layers from the stomach to the anus is fairly similar and contains the following types of tissue:

1. *Mucosa:* the inner epithelial tissue lining of the alimentary canal, consisting of simple columnar epithelium. This layer contains *glands* that secrete products such as hydrochloric acid, enzymes, and mucus, as well as a collection of lymphoid nodules called *mucosal associated lymphoid tissue* (MALT). In the ileum, the lymphoid nodules are termed *Peyer's patches.* The mucosa also houses a very thin layer of smooth muscle called the *muscularis mucosa.*

2. *Submucosa*: a layer of connective tissue that houses blood vessels, nerves, and lymphatics, as well as elastic fibers.

3. *Muscularis externa*: two layers of smooth muscle—an inner circular layer and an outer longitudinal layer. It produces the rhythmic contractions of peristalsis by alternating contraction of the layers of smooth muscle.

4. *Serosa*: the outer connective tissue layer; functionally, the visceral peritoneum. It is also known as the *adventitia*.

The esophagus has some notable differences from the rest of the alimentary canal, as follows:

1. *Mucosa*: The mucosa of the esophagus is composed of *stratified squamous epithelium*. The esophagus lacks the thick layer of mucus that protects other areas of the alimentary canal. As a result, it requires a thicker, tougher epithelial tissue. At the junction of the esophagus and stomach, you are able to see an abrupt transition from stratified squamous epithelium to simple columnar epithelium.

2. *Muscularis externa*: The muscularis externa of the esophagus changes as it progresses toward the stomach.
   a. The upper 1/3 of the esophagus is *skeletal muscle*.
   b. The middle 1/3 of the esophagus is about *half skeletal muscle and half smooth muscle*.
   c. The lower 1/3 of the esophagus is *smooth muscle*, and is controlled by the nervous system.

These differences in the muscularis externa allow you to determine the location of the section you are examining. Differences are also visible in the stomach:

1. *Mucosa*: The mucosa of the stomach is heavily indented, reflecting the presence of numerous *gastric glands*. These glands secrete products for digestion known collectively as *gastric juice*. The mucosa of the stomach also contains a large number of mucus-secreting cells known as *goblet cells*. Goblet cells are found throughout the alimentary canal but are concentrated in the stomach.

2. *Muscularis externa*: Whereas the majority of the alimentary canal contains only two layers of smooth muscle in the muscularis externa, the stomach contains three layers —an inner circular, a *middle oblique*, and an outer longitudinal layer of smooth muscle. This allows the stomach to undergo a motion known as *churning*, which functions to mechanically pummel the food into a liquid mixture called *chyme*.

In this exercise you also will be examining three examples of accessory organs—a tooth, the liver, and the pancreas. In the pancreas, you will see two different types of cells:

1. *Acinar cells*: the exocrine cells of the pancreas. They secrete enzymes, bicarbonate, and water into ducts that merge to form the *main pancreatic duct*, which empties into the duodenum.

2. *Islets of Langerhans* (also called *pancreatic islets*): the endocrine cells of the pancreas that secrete hormones such as insulin into the bloodstream. They are visible as small, circular groups of cells that stain less darkly than the surrounding acinar cells. We will reexamine these groups of cells in Unit 26, the Endocrine System.

In the liver you will see the following features:

1. *Hexagonal liver lobules*: Hepatocytes (liver cells) are arranged into hexagonal plates of cells called liver lobules. At the center of each lobule is a central vein, which eventually will drain into the hepatic veins.

2. *Portal triads*: areas at the corners of the liver lobules that house three small vessels:
   a. *Bile duct*: carries bile made by hepatocytes and drains into the hepatic duct.
   b. *Portal vein (venule)*: a tiny branch off the hepatic portal vein that delivers nutrient-rich blood to the liver for processing and detoxification.
   c. *Hepatic artery (arteriole)*: delivers oxygen-rich blood to nourish the hepatocytes.

Models

Identify the following histological structures on a model of the four layers of the alimentary canal before moving on to prepared slides.

1. Mucosa
   a. Epithelial tissue (simple columnar epithelium)
   b. Glands
   c. MALT
   d. Muscularis mucosae

2. Submucosa
   a. Blood vessels
   b. Nerves
   c. Lymphatic vessels

3. Muscularis externa
   a. Inner circular layer
   b. Outer longitudinal layer

4. Serosa

Slides

Examine the following prepared microscope slides. Use colored pencils to draw what you see, and label the structures indicated below. For the majority of the slides, it is useful to begin on low power to gain a full view of all histologic layers, and then move up to higher power to identify specific structures on the slide. The figure numbers refer to figures in your lab atlas.

Human or Mammal Tooth

(See Figures 15.7 and 15.8 in your lab atlas.)

Label the following on your drawing:

1. Enamel (may not be visible)
2. Dentin
3. Pulp cavity
4. Gingiva
5. Alveolar bone

Esophagus

(See Figures 15.14 and 15.15 in your lab atlas.)

Label the following on your drawing:

1. Stratified squamous epithelium
2. Submucosa with blood vessels
3. Muscularis externa
   a. Smooth muscle
   b. Skeletal muscle

Gastroesophageal Junction

(See Figure 15.23 in your lab atlas.)

Identify the abrupt change of epithelium from stratified squamous epithelium lining the esophagus to simple columnar epithelium lining the stomach.

Fundus of the Stomach

(See Figure 15.24 in your lab atlas.)

Label the following on your drawing:

1. Mucosa with goblet cells and gastric glands
2. Submucosa with blood vessels
3. Muscularis externa with smooth muscle layers:
   a. Inner circular
   b. Middle oblique
   c. Outer longitudinal
4. Serosa

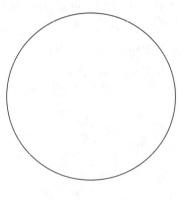

Duodenum

(See Figure 15.31–15.35 in your lab atlas.)

Label the following on your drawing:

1. Villi
2. Mucosal-associated lymphoid tissue
3. Microvilli (visible as small "hairs" projecting from the cells)
4. Lacteal
5. Intestinal glands

Ileum

Label the following on your drawing: (There is no diagram of the ileum in your lab atlas, but see Figure 13.7—the lymphatic nodules in the tonsil resemble Peyer's patches.)

1. Mucosa

2. Submucosa
   a. Peyer's patches

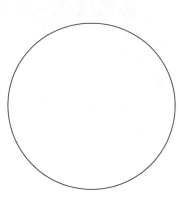

Pancreas

(See Figures 10.11 and 15.52 in your lab atlas.)

Label the following on your drawing:

1. Acinar cells

2. Pancreatic ducts

3. Pancreatic islets
   (islets of Langerhans)

Liver

(See Figures 15.46–15.48 in your lab atlas.)

Label the following on your drawing:

1. Liver lobule

2. Portal triad

3. Central vein

✔ Check Your Understanding

1. The small intestine has three progressively smaller folds of its lining—the plicae circulares, the villi, and the microvilli. Considering the primary function of the small intestine, explain why the small intestine has these folds.

2. The condition known as heartburn is most often caused by acid regurgitating from the stomach into the esophagus. Why do you think the acid tends to burn the esophagus and produce pain but does not similarly burn the stomach?

3 EXERCISE 3: Tracing Exercises

In this exercise you will be tracing the pathway that three different nutrients take from ingestion at the mouth to their arrival at the heart. You will trace a cookie (primarily carbohydrates), an egg (primarily protein), and a greasy French fry (primarily lipids).

Along the way, detail the following for each:

1. The *anatomical pathway* that each takes, from ingestion through its passage through the alimentary canal, to its absorption into the blood, and finally to its travels through the blood until it reaches the heart.

2. The *enzymes* that chemically digest each substance at different locations, as well as any physical processes that serve to break the food into smaller pieces (churning, chewing, emulsification, etc.).

Some hints:

● Don't forget about the hepatic portal system!

● Remember that the process of digestion and absorption is quite different for lipids. For example, fats are not absorbed into the same structures as proteins and carbohydrates, but instead are absorbed into a specialized structure.

● Use your list of enzymes that you completed in the Pre-Lab Exercises for reference.

● Refer to the tracing exercises from Unit 17, Blood Vessel Anatomy, to review the pathway of blood flow through the body.

1. Cookie:

2. Egg:

3. Greasy French fry:

✔ **Check Your Understanding**

1. Why are the processes for digestion and absorption of lipids so much more complex than those for carbohydrates and proteins? (*Hint*: think about the types of chemical bonds in lipids.)

2. Why are lipids absorbed into a different structure than carbohydrates and proteins?

4. Add three or four drops of red Sudan stain, and shake the tube again for 15 seconds. What color is the oil? What color is the water?

4 EXERCISE 4: Observing Emulsification in Action _____

In the Pre-Lab Exercises you learned about emulsification and how emulsification of lipids in the digestive system is accomplished by bile salts. In this exercise we will observe emulsification in action. You will use four things:

1. *Lipids*: The source of lipids for this exercise is vegetable oil.

2. *Emulsifying agent*: Detergents are considered to be emulsifiers because they have both polar parts and nonpolar parts, similar to bile salts. The emulsifying agent in this exercise, therefore, will be a liquid detergent.

3. *Distilled water.*

4. *Red Sudan stain*: This is a stain that binds only to lipids. The only purpose of this stain is to make the lipids more visible during the procedure.

5. Add about 1 mL of the liquid detergent to the mixture, and shake the tube vigorously again. Allow the tube to stand again. What has happened to the solution? Is it still two distinct colors? Explain your results.

Procedure

> NOTE: SAFETY GLASSES AND GLOVES ARE REQUIRED.

1. Fill a test tube about one-fourth full of distilled water (approximately 2 mL in a standard test tube).

2. Add about the same amount of vegetable oil to the water.

3. Cover the tube and shake it vigorously for 15 seconds. Allow it to stand for 2 minutes. What happens to the oil and water?

26. Endocrine System

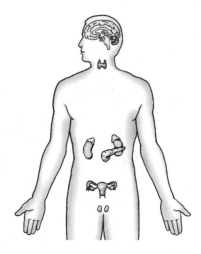

■ **OBJECTIVES**

Once you have completed this unit, you should be able to:

1. Identify endocrine organs and structures on anatomical models and preserved specimens.

2. Observe the histology of endocrine organs on microscope slides.

3. Describe the functions, stimulus for secretion, and target tissues of various hormones.

4. Solve clinical endocrine "mystery cases."

■ **MATERIALS**

● Anatomical models: endocrine system chart, torsos, head and neck

● Microscope slides: pancreas, thyroid gland, pituitary gland, adrenal gland

● Colored pencils

● Fetal pigs or other preserved small mammal

● Dissection equipment

PRE-LAB EXERCISES

Prior to coming to lab, complete the following exercises, using Chapter 10 in your lab atlas for reference.

1 **PRE-LAB EXERCISE 1:**
Endocrine Organs
and Hormones

The endocrine system consists of a diverse set of organs, all of which secrete chemicals called *hormones* into the bloodstream. Fill in the accompanying chart with the locations and hormones secreted by each of the endocrine organs listed.

2 **PRE-LAB EXERCISE 2:**
Endocrine System
Anatomy _____

Label Figure 26.1, depicting the endocrine system, using the terms from Exercise 1.

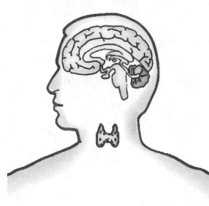

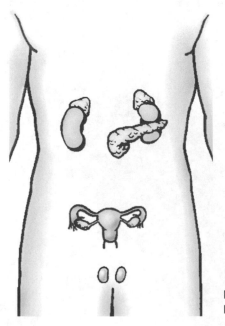

ENDOCRINE ORGAN	LOCATION	HORMONE(S) SECRETED
Hypothalamus		
Anterior pituitary gland		
Posterior pituitary gland		
Pineal gland		
Thyroid gland		
Thymus gland		
Parathyroid glands		
Pancreas		
Adrenal cortex		
Adrenal medulla		
Testes		
Ovaries		

FIGURE 26.1
Diagram of the endocrine system.

3 PRE-LAB EXERCISE 3:
Hormones' Target Tissues and Effects

Each of the hormones you listed in the Pre-Lab Exercise 1 acts on one or more tissues (its *target tissue*) to produce a specific effect. For each of the hormones listed in the following chart, list the hormone's target tissue(s) and the effects the hormone has on that tissue.

HORMONE	TARGET TISSUE(S)	HORMONE EFFECTS
Antidiuretic hormone		
Oxytocin		
Thyroid-stimulating hormone		
Adrenocorticotropic hormone		
Growth hormone		
Follicle-stimulating hormone		
Lutenizing hormone		
Prolactin		
Melatonin		
Thyroxine and triiodothyronine (T4 and T3)		
Calcitonin		
Thymosin and thymopoietin		
Parathyroid hormone		
Cortisol		
Aldosterone		
Estrogen		
Testosterone		
Progesterone		
Epinephrine and norepinephrine		
Insulin		
Glucagon		

EXERCISES

The endocrine system, a diverse group of ductless glands, is one of the major homeostatic systems in the body. The other major system that acts to maintain homeostasis, which we have already discussed, is the nervous system. The method by which these two systems maintain homeostasis is different, though. The nervous system works via nerve impulses to produce effects that are nearly immediate, but temporary. The endocrine system works via secretion of *hormones*—chemicals secreted into the bloodstream that typically act on distant targets. The actions of hormones are typically not immediate, but the effects are longer-lasting.

This unit introduces you to the anatomy and histology of the endocrine organs. To close out this unit, you will play "endocrine detective" and try to solve three "endocrine mysteries."

▌ EXERCISE 1:
Endocrine System Anatomy_____

In general, hormones function to regulate the processes of other cells, including inducing the production of enzymes or other hormones, changing the metabolic rate of the cell, and altering the permeability of cell membranes. I often liken hormones to the "middle managers" of the body, because they communicate the messages from their "bosses" (the endocrine glands) and tell other cells what to do. Some endocrine glands secrete hormones as their primary function (e.g., the thyroid and anterior pituitary glands). Others, however, secrete hormones as a secondary function. Examples include the heart (atrial natriuretic peptide), adipose tissue (leptin), kidneys (erythropoietin), and stomach (gastrin).

Identify each of the following endocrine organs on anatomical models and diagrams. You may wish to use a fetal pig or other preserved small mammal to identify some of the organs, such as the thymus, that may not be visible on the models. See Pre-Lab Exercise 1 and Figure 10.1 in your atlas for reference.

Endocrine Glands

1. Hypothalamus
   a. Infundibulum
2. Pituitary gland
   a. Anterior pituitary
   b. Posterior pituitary
3. Pineal gland
4. Thyroid gland
   a. Isthmus
5. Parathyroid glands
6. Thymus gland
7. Adrenal glands
   a. Adrenal cortex
   b. Adrenal medulla
8. Pancreas
9. Ovaries
10. Testes

Model Inventory

As you examine the anatomical models and diagrams in lab, list them on the inventory below, and detail which structures you are able to locate on each model.

MODEL/DIAGRAM	STRUCTURES IDENTIFIED

✔ Check Your Understanding

1. In adults, the thymus gland atrophies to become mostly adipose and other connective tissue. Explain this, considering the function of the thymus gland.

2. The adrenal medulla increases its output of catecholamines (epinephrine and norepinephrine) in response to stimulation by the sympathetic nervous system. During times of sympathetic nervous system activity, however, secretions from the adrenal cortex also increase. How might the hormones of the adrenal cortex help the body adapt to stress?

3. Tumors of the parathyroid gland often result in secretion of excess parathyroid hormone. Considering the function of this hormone, predict the effects of such a tumor.

2 EXERCISE 2: Histology of Endocrine Organs

In this exercise we will examine the histology of four endocrine organs: the thyroid gland, the adrenal gland, the pituitary glands, and the pancreas (we will examine the ovary and testis in the next unit).

1. *Thyroid gland*: Microscopically, the thyroid gland is composed of structural units called *follicles*. Follicles are spherical, hollow structures lined by a ring of cuboidal *follicle cells* that produce *thyroid hormone*. Inside the follicles is the reddish-brown, iodine-containing substance called *colloid*. As the follicle cells make thyroid hormone, it enters the colloid, where iodine residues are added to it. In between the follicles are clusters of cells called *parafollicular cells* (also called *C cells*). These cells produce the hormone *calcitonin*, which works with parathyroid hormone to maintain calcium balance in the blood. The features of thyroid follicles are best viewed on high power.

2. *Adrenal gland*: The adrenal gland is composed of two main regions: the outer *adrenal cortex*, which secretes *steroid* hormones, and the inner *adrenal medulla*, which secretes epinephrine and norepinephrine in response to sympathetic nervous system stimulation. The adrenal cortex may be further subdivided into three zones: the outer *zona glomerulosa*, a thin layer where the cells are arranged in clusters; the middle *zona fasciculata*, where the cells are stacked on top of one another and resemble columns; and the inner *zona reticularis*, where the cells stain more darkly and are tightly packed.

   The innermost adrenal medulla is distinguished from the zones of the cortex because it contains numerous blood vessels with loosely arranged cells. The regions of the adrenal gland are best viewed on low or medium power. To get more detail of the individual cells, switch to high power.

3. *Pituitary gland*: The two divisions of the pituitary gland are composed of two different types of tissue: The anterior pituitary is glandular epithelium, and the posterior is neural tissue (hence their other names, the *adenohypophysis* and *neurohypophysis*, respectively). The posterior pituitary is composed of axons, which appear fine and granular, and scattered cells called *pituicytes*. The anterior pituitary contains three types of cells that can be distinguished based upon their staining properties. The reddish-brown *acidophiles* produce prolactin and growth hormone, and the bluish-purple *basophils* produce thyroid-stimulating hormone, adrenocorticotropic hormone, follicle-stimulating hormone, and lutenizing hormone. The third population of cells, which do not take up any stain and therefore are called *chromophobes*, have an unknown function.

   Begin your examination of the pituitary gland on low power so you can see the basic differences between the anterior and posterior pituitary. Then, move up to high power to differentiate the three types of cells in the anterior pituitary.

4. *Pancreas*: The pancreas is a gland that has both endocrine and exocrine functions. In Unit 25 you looked at the exocrine portion of the pancreas, the enzyme-secreting *acinar cells*. Most of the pancreas is composed of acinar cells; however, as you scan the slide on low or medium power, you will note small "islands" of tissue.

These islands are the insulin- and glucagon-secreting *islets of Langerhans*, or simply *pancreatic islets*. Typically, the islets are lighter in color than the surrounding acinar cells. On high power, the differences between the two tissues are easily seen: The acinar cells are slightly cuboidal cells arranged around a duct, whereas the islet cells have no distinctive arrangement (there also may be larger, stain-free spaces between the islet cells).

Slides

Examine each of the following prepared slides of endocrine glands. Use colored pencils to draw what you see, and label your drawing with the structures indicated below. The references to figures refer to figures in your lab atlas.

Thyroid

(See Figure 10.6 in your lab atlas.)

1. Follicle cells
2. Colloid
3. Parafollicular (C) cells

Pituitary

(See Figures 10.2–10.4 in your lab atlas.)

1. Anterior pituitary (adenohypophysis)
2. Posterior pituitary (neurohypophysis)

3. Anterior pituitary (on high power)
   a. Acidophiles
   b. Basophils
   c. Chromophobes

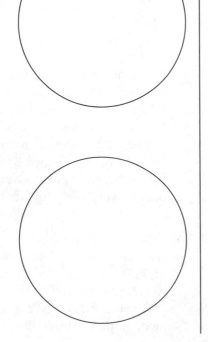

Adrenal Gland

(See Figures 10.8–10.10 in your lab atlas.)

1. Zones of the adrenal cortex:
   a. Zona glomerulosa
   b. Zona fasciculata
   c. Zona reticularis
2. Adrenal medulla

Pancreas

(See Figure 10.11 in your lab atlas.)

1. Acinar cells (exocrine portion)
2. Islets of Langerhans

✔ Check Your Understanding

1. The hormone *calcitonin* is available commercially as a nasal spray prescribed to treat osteoporosis. Explain why this hormone would help to reduce bone loss, and in some cases build bone, in patients affected with this disease.

2. The drug spironolactone is part of a group of drugs known as *potassium-sparing diuretics*. Spironolactone works by blocking the effects of aldosterone on the kidney. What effect would this class of drugs have on urine output, and consequently on blood pressure? Why would these drugs spare potassium (i.e., prevent excess potassium from being excreted)? Could these drugs have a negative impact on the acid–base balance of the body?

3. The disease *diabetes mellitus, type I* is characterized by destruction of the cells that produce insulin in the islets of Langerhans. Considering the functions of insulin, predict the symptoms of this disease.

3 EXERCISE 3: Endocrine Mystery Cases _____

In this exercise you will be playing the role of "endocrine detective" to solve various endocrine mysteries. In each case you will have a victim who has suddenly fallen ill with a mysterious malady. You will be presented with a set of "witnesses," each of whom will give you a clue as to the nature of the illness. Other clues will come from samples that you will send off to the lab for analysis. You will solve the mystery by providing the victim with a diagnosis.

Case 1: The Cold Colonel

You are called upon to visit the ailing Col. Lemon. Before you see him, you speak with three witnesses who were with him when he fell ill.

Witness statements:

- *Ms. Magenta*: "Colonel Lemon has been hot-blooded for as long as I've known him. But I noticed that he couldn't seem to keep warm. He kept complaining about being cold. . . ."
- *Mr. Olive*: "Just between you and me, I've noticed that the old chap has put on quite a bit of weight lately."
- *Professor Purple*: "The Colonel and I used to go on major expeditions together. Now he just doesn't seem to have the energy to do much of anything."

What are your initial thoughts about the witnesses' statements? Does one hormone come to mind that may be the cause?

You see the Colonel and collect some blood to send off to the lab. The analysis comes back as follows:

T3 (triiodothyronine): 50 ng/dl (normal: 110–230 ng/dl)

T4 (thyroxine): 1.1µg/dl (normal: 4–11 µg/dl)

TSH (thyroid-stimulating hormone): 86 µU/ml (normal: 2–10 µU/ml)

Analyze the results. Why are the T3 and T4 low and the TSH high?

Based upon the witness statements and the laboratory analysis, what is your final diagnosis? Explain.

Case 2: The Bloated Ms. Blanc

Your next call is to the home of Ms. Blanc. As before, you have three witnesses to interview.

Witness statements:

- *Ms. Magenta*: "Between you and me, she has really let herself go. She has fat deposits in all of these strange places, like around her face and trunk, and a weird hump on her back. If you ask me, some exercise may do her some good!"

- *Col. Lemon*: "Not that I want to talk about such things, but she has seemed bloated, a bit swollen, lately. She told me that even her blood pressure has gone up!"

- *Mrs. Feather*: "The poor dear. She has been sick so much lately with all kinds of infections. Personally, I think her immune system needs boosting. But my herbal teas don't seem to help."

What are your initial thoughts, based upon the witnesses' statements? Does one hormone come to mind that could produce these effects?

Your next stop is to speak with Ms. Blanc. During the interview you notice a bottle of pills on her nightstand. You log the pills as evidence and send them off to the lab for analysis. The lab report shows the pills to be the medication *prednisone*, which you know to be a glucocorticoid similar to the hormone cortisol taken to reduce inflammation. Is this finding significant? Why or why not?

Based upon the witness statements and the laboratory analysis, what is your final diagnosis? Explain.

Case 3: The Parched Professor

Your last call is to the aid of Professor Purple. Three witnesses are present from whom to take statements.

Witness statements:

- *Mr. Olive*: "I swear that I saw him drink a full glass of water every half an hour today. He kept saying how thirsty he was!"

- *Ms. Blanc*: "He must be going to the . . . well, you know, the little boys' room, two or three times every hour!"

- *Mrs. Feather*: "He has been saying lately that his mouth is dry and that he feels weak. Personally, I think he's just not following a healthy diet! He should be drinking some of my herbal teas!"

Based upon the witnesses' statements, what are your initial thoughts? Does one hormone come to mind that could produce these effects?

You interview Professor Purple and collect blood and urine specimens to be sent off to the lab for analysis. The lab reports that the urine osmolality is 150 mOsm/kg, which means that the urine is overly dilute; and that the blood osmolality is 300mOsm/kg, meaning that the blood is overly concentrated. What is the significance of these clues?

Based upon the witness statements and the laboratory analysis, what is your final diagnosis? (*Hint*: Think of the hormone that is supposed to trigger water retention from the kidneys. Is there a disease where this hormone is deficient?)

27. Reproductive System

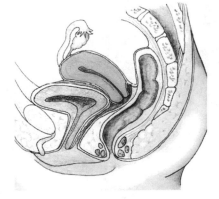

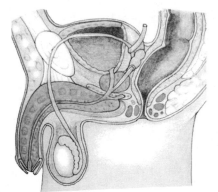

■ OBJECTIVES

Once you have completed this unit, you should be able to:

1. Identify structures of the male and female reproductive system on anatomical models.

2. Observe the microscopic anatomy of male and female reproductive structures on microscope slides.

■ MATERIALS

● Anatomical models: male reproductive system, female reproductive system

● Microscope slides: testes, fallopian tube, epididymis

● Colored pencils

PRE-LAB EXERCISES

Prior to coming to lab, complete the following exercises, using Chapter 17 in your lab atlas for reference.

PRE-LAB EXERCISE 1:
Male Reproductive Anatomy

Label and color-code Figure 27.1, illustrating the male reproductive tract, using the terms from Exercise 1. Please note that all structures may not be visible on the diagrams.

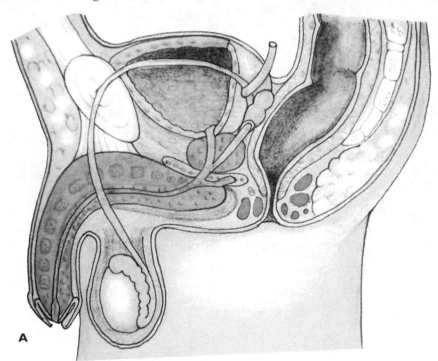

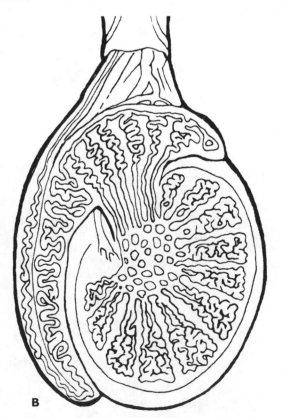

FIGURE 27.1
Male reproductive tract: (A) Sagittal section of the male pelvis; (B) Section through the testis.

2 PRE-LAB EXERCISE 2:
Female Reproductive Anatomy

Label and color-code Figure 27.2, illustrating the female reproductive tract, using the terms from Exercise 2. Please note that all listed structures may not be visible on the diagrams.

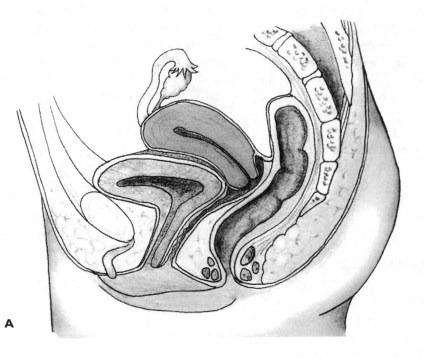

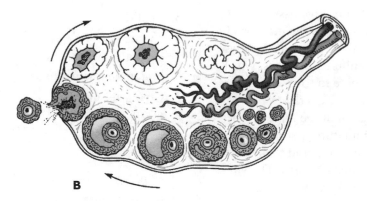

FIGURE 27.2
Female reproductive tract: (A) Sagittal section of the female pelvis; (B) Section through the ovary.

EXERCISES

The other organ systems in the human body that we have discussed all function in some manner to help maintain homeostasis of the body. The reproductive system is unique, however, in that it plays little role in maintenance of homeostasis, and instead functions to perpetuate the species. The main organs of the reproductive system are the *gonads*, the testes and ovaries, which produce *gametes*, or sex cells, for reproduction.

The exercises in this unit relate to the structures of the male (Exercise 1) and female (Exercise 2) reproductive anatomy. You will also examine slides of reproductive tissue and identify histologic features of the gonads and other reproductive organs.

1 EXERCISE 1: Male Reproductive Anatomy _____

The *testes*, the gamete-producing organs of the male, are situated outside the body in a sac of skin and connective tissue called the *scrotum*. Recall from Unit 26 that the testes also are endocrine organs, and produce the hormone testosterone. The testes lie outside the body because sperm production will not take place at body temperature, instead requiring a temperature of about 34 degrees C (about 94 degrees F).

The testes are surrounded by a connective tissue sheath called the *tunica albuginea*, which dives into the interior of the testes to form lobules, each of which contains a tightly coiled *seminiferous tubule*. The seminiferous tubules, which are the site of *spermatogenesis*, converge to form a structure called the *rete testis*. The rete testis then exits the testis to join the first segment of the duct system of the male reproductive tract, the *epididymis*. The immature sperm migrate to the epididymis to finish their maturation, and then exit via a long tube called the *vas* (or ductus) *deferens*. The vas deferens travels superiorly through the *spermatic cord*, a structure that also carries the testicular artery, testicular veins (the pampiniform plexus), and nerves.

Once the vas deferens enters the pelvic cavity, it crosses superiorly and posteriorly over the bladder, where it joins with a gland called the *seminal vesicle* and forms the *ejaculatory duct*. This duct passes through the *prostate gland*, where it joins with the *prostatic urethra*. The prostatic urethra becomes the *membranous urethra* as it exits the prostate, and then becomes the *spongy urethra* as it enters the corpus spongiosum of the penis.

The male reproductive tract consists of three exocrine glands: the prostate, seminal vesicles, and *bulbourethral glands*. Both the seminal vesicles and the prostate produce about 90% of the volume of *semen*, a fluid that contains

chemicals to nourish and activate the sperm. The smaller bulbourethral glands produce an alkaline secretion that is released prior to the release of sperm during an ejaculation. Because the urethra also transports urine, the pH in the urethra is generally acidic. The alkaline fluid neutralizes the acid in the urethra, which would inactivate the sperm.

The *penis* is composed of three erectile bodies: the single *corpus spongiosum* and the paired, dorsal *corpora cavernosa*. The corpus spongiosum, which surrounds the spongy urethra, enlarges distally to form the *glans penis*. All three bodies consist of vascular spaces that fill with blood during an erection.

Identify the following structures of male reproductive anatomy on models and diagrams. Use Figures 17.3–17.5, 17.17, and 17.18 in your lab atlas and Pre-Lab Exercise 2 for reference.

Male Reproductive Anatomy

1. Scrotum
2. Testes
   a. Tunica albuginea
   b. Tunica vaginalis
   c. Seminiferous tubules
   d. Rete testis
3. Epididymis
4. Spermatic cord
   a. Testicular arteries
   b. Pampiniform plexus
5. Cremaster muscle
6. Vas deferens
7. Ejaculatory duct
8. Urethra
   a. Prostatic urethra
   b. Membranous urethra
   c. Spongy urethra
   d. External urethral orifice
9. Glands:
   a. Seminal vesicle
   b. Prostate gland
   c. Bulbourethral gland
10. Penis
    a. Corpora cavernosa
    b. Corpus spongiosum
       (1) Glans penis

Model Inventory

As you examine the anatomical models and diagrams in lab, list them on the inventory below, and detail which structures you are able to locate on each model.

MODEL/DIAGRAM	STRUCTURES IDENTIFIED

✔ Check Your Understanding

1. A condition called *testicular torsion* results when the spermatic cord becomes twisted. Why is this considered a surgical emergency?

2. What structure is sectioned in a vasectomy? What are the effects of this procedure? Does it impact sperm production? Will it significantly impact the volume of semen produced?

3. The condition *benign prostatic hypertrophy*, in which the prostate is enlarged, often results in urinary retention —the inability to completely empty the bladder. Considering the anatomy of the male genitourinary tract, explain this symptom.

4. Explain why men with a low sperm count often are advised not to wear briefs, and to avoid excessive time in hot tubs.

EXERCISE 2:
Female Reproductive Anatomy_____

The female reproductive organs lie in the pelvic cavity, with the exception of the almond-shaped *ovaries*, which are located in the peritoneal cavity. The ovaries, the female gonads, produce gametes (*oocytes*, or eggs), which travel through the reproductive tract to be fertilized. Recall that the ovaries, like the testes, are endocrine glands, and they produce the hormone *estrogen*. The ovaries are held in place by several ligaments, including the *ovarian ligament*, which anchors them to the uterus (this is sometimes confused by students as the uterine tube), a part of the *broad ligament* called the *mesovarium*, and the *suspensory ligament*. The suspensory ligaments extend from the lateral body wall and carry with them the blood supply of the ovary.

Within the ovaries, developing oocytes are encased in structures called *follicles*. Follicles are present in various stages, ranging from the immature *primordial follicles* to the mature *Graafian* (vesicular) *follicles*. At the stage of the vesicular follicle, the ovary wall ruptures and the oocyte is released. Unlike the male reproductive tract, in which the tubule system is continuous, the tubule system of the female reproductive tract is not continuous. Therefore, the oocyte is released not directly into the uterine tube but, instead, into the pelvic cavity. It is the job of the fingerlike extensions of the *uterine* (fallopian) *tube* called *fimbriae* to "catch" the ovulated oocyte and bring it into the uterine tube.

The uterine tubes join the superolateral portion of the *uterus*. The uterus, a small organ that lies between the urinary bladder and the rectum, consists of three portions: the dome-shaped *fundus*, the *body*, and the narrow *cervix*. The wall of the uterus has three layers: the inner epithelial and connective tissue lining called the *endometrium*, in which a fertilized ovum implants; the muscular *myometrium*; and the outermost connective tissue lining, the *perimetrium* (which is actually just the visceral peritoneum).

From the inferiormost portion of the cervix, called the *cervical os*, extends the *vagina*. Serving both as the birth canal and as the copulatory organ, the vagina is approximately 4 inches long and terminates in the vaginal orifice. Flanking the vaginal orifice are the *greater vestibular* (Bartholin's) *glands*, which secrete mucus to lubricate the vaginal canal during coitus.

The external anatomy of the female is collectively called the *vulva*, which consists of the *labia majora* and *minora*, the *clitoris*, and the urethral and vaginal orifices. The labia are paired skinfolds that are analogous to the scrotum in the male. They enclose a region called the *vestibule*, which houses the clitoris, as well as the urethral and vaginal orifices.

The *mammary glands*, though not true reproductive organs, have an associate reproductive function in milk production to nourish the newborn infant. Mammary glands are present in both males and females (males can produce milk too), but their anatomy is most appropriately discussed with female anatomy. Internally, mammary glands consist of 15–25 *lobes*, each of which has smaller *lobules* that contain milk-producing *alveoli*. Milk leaves the alveoli through *lactiferous ducts*, which join to form storage areas called *lactiferous sinuses*. Milk leaves through the *nipple*, which is surrounded by a darkly pigmented area called the *areola*.

Identify the following structures of the female reproductive system on models and diagrams using your textbook and Figures 17.19–17.21, 17.23, 17.28, and 17.34–17.35 in your lab atlas and Pre-Lab Exercise 1 for reference.

Female Reproductive Anatomy

1. Ovary
2. Ovarian follicles
   a. Primordial follicle
   b. Primary follicle
   c. Secondary follicle
   d. Graafian (vesicular) follicle
3. Uterine (fallopian) tube
   a. Fimbriae
4. Uterus
   a. Fundus
   b. Body
   c. Cervix
   d. Layers
      (1) Endometrium
      (2) Myometrium
      (3) Perimetrium
5. Vagina

6. Vulva
   a. Vestibule
   b. Labia majora
   c. Labia minora
   d. Clitoris
   e. Urethral orifice

7. Ligaments
   a. Broad ligament
     (1) Mesometrium
     (2) Mesovarium
     (3) Mesosalpinx
   b. Round ligament
   c. Suspensory ligament of the ovary
   d. Ovarian ligament

8. Mammary glands
   a. Areola
   b. Nipple
   c. Lobe
   d. Lobule (with alveoli)
   e. Lactiferous duct
   f. Lactiferous sinus
   g. Adipose tissue

Model Inventory

As you examine the anatomical models and diagrams in lab, list them on the inventory below, and detail which structures you are able to locate on each model.

Model / Diagram	Structures Identified

✔ Check Your Understanding

1. Why might *pelvic inflammatory disease*, a condition usually caused by bacterial infections, result in symptoms outside the female reproductive tract? (*Hint*: Think about the lack of a direct connection between the ovaries and the uterine tubes.)

2. If a fertilized ovum implants in the wall of the uterine tube, this results in a tubal (or ectopic) pregnancy. Why is this dangerous? What may increase a woman's risk for an ectopic pregnancy?

3. One of the most common complaints of pregnant women is the need to urinate often. Explain this, considering the anatomy of the organs in the pelvic cavity.

3 EXERCISE 3:
Reproductive Histology_____

In this exercise we will examine the histology of the testes, epididymis, and fallopian tube. We will examine the ovary in Unit 28. Following are some hints about what to look for as you examine the slides.

- *Testes*: The testes likely will show sections of the seminiferous tubules. On low power, you will see several small, roughly circular, sections, which are cross-sections of the tubules. You also may be able to see the tunica albuginea and the rete testis (see Figure 17.6 in your atlas). Advance to higher power to see the tubules in better detail (my students have remarked that they look like kiwi fruit). On the outermost rim of the tubule, you will see the cuboidal *spermatogonia* (the earliest stage of sperm cells). As you move toward the lumen, the cells get progressively smaller until they are mere packages of DNA. These are the *spermatids*, which will migrate from here to the epididymis to mature. In between the tubules, you will see small clusters of cells called *interstitial* (or Leydig) *cells*. These cells make testosterone, which is required for spermatogenesis to take place.

- *Epididymis*: The epididymis is lined with pseudostratified columnar epithelium. The cells contain projections called *stereocilia*, which are not really cilia but, instead, long microvilli that function to nourish the sperm as they mature. On high power, you should be able to see the stereocilia and also sperm in the lumen of the tubule.

- *Uterine tube*: A cross-section of the uterine tube reveals a highly convoluted mucosal surface that is lined with ciliated simple columnar epithelium. Examine the section first on low power to see the convoluted surface, and then advance to high power to see the epithelium with the cilia.

Examine each of the following prepared slides of reproductive structures. Use colored pencils to draw what you see, and label your drawing with the names of the structures indicated below. The figures referenced are in your lab atlas.

Testes
(See Figures 17.6 and 17.8–17.10 in your lab atlas.)

1. Rete testis
2. Tunica albuginea

3. Seminiferous tubules (on high power)
   a. Lumen
   b. Spermatids
   c. Spermatogonia
   d. Interstitial cells

Epididymis
(See Figures 17.11–17.12 in your lab atlas.)

1. Pseudostratified columnar epithelium
2. Stereocilia
3. Sperm

Uterine (Fallopian) Tube

1. Lumen
2. Simple columnar epithelium
3. Cilia

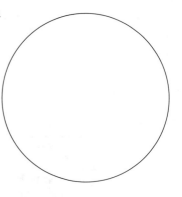

28. Gametogenesis and Human Development

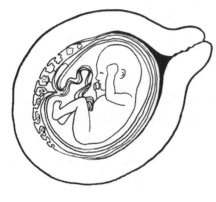

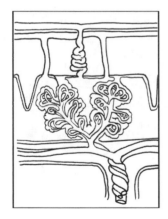

■ OBJECTIVES

Once you have completed this unit, you should be able to:

1. Identify the stages of meiosis, and describe how it is different from mitosis.

2. Examine and describe the histology of the ovary, seminiferous tubules, and spermatozoa.

3. Describe the processes of fertilization and implantation.

4. Identify and describe structures and membranes associated with the fetus.

5. Identify and describe the structures of fetal circulation.

6. Trace a developing embryo from fertilization and implantation, and through the stages of embryonic and fetal development.

■ MATERIALS

- Anatomical models: meiosis models, mitosis models (for comparison), sequence of embryonic development, female reproductive system with fetus, fetal circulation

- Microscope slides: ovary, seminiferous tubules, and sperm smear

- Multicolored pipe cleaners

- Colored pencils

PRE-LAB EXERCISES

Prior to coming to lab, complete the following exercises, using Chapters 2 and 18 in your atlas for reference.

1 PRE-LAB EXERCISE 1:
Ovary and Ovarian Follicles_____

Label and color-code the diagram of the ovary and ovarian follicles illustrated in Figure 28.1, using the terms from Exercise 2.

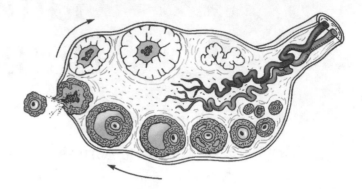

FIGURE 28.1
Diagram of ovary.

2 PRE-LAB EXERCISE 2:
Membranes and Structures Surrounding the Fetus_____

Label and color-code the membranes and structures surrounding the fetus using the terms from Exercise 3. Please note that not all structures may be visible in the diagram.

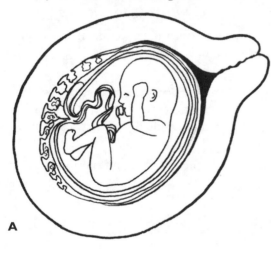

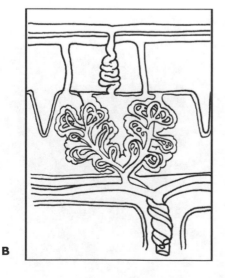

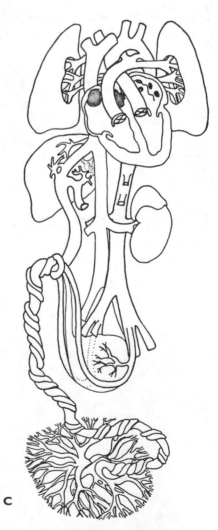

FIGURE 28.2
(A) Diagram of fetus in uterus with membranes, placenta, vessels; (B) Diagram of chorionic villi; (C) Diagram of fetal heart and circulation.

3 PRE-LAB EXERCISE 3:
Embryonic and Fetal Development _____

Define the following terms associated with embryonic and fetal development:

TERM	DEFINITION
Stages of development	
Zygote	
Morula	
Blastocyst	
Implanting blastocyst	
Embryo	
Fetus	
Fetal structures	
Chorion	
Chorionic villi	
Amnion	
Placenta	
Yolk sac	

4 PRE-LAB EXERCISE 4:
Stages of Mitosis _____

In this unit we will discuss the process of meiosis. Meiosis proceeds in a similar fashion to mitosis, necessitating a review of mitosis before we begin the lab. Describe the events that are occurring in the cell in each of the following stages of mitosis:

STAGE OF MITOSIS	EVENTS OCCURRING IN THE CELL
Prophase	
Metaphase	
Anaphase	
Telophase	

EXERCISES

Gametogenesis is the production of ova and sperm by the gonads. It is a unique process in that the end results—gametes—have half the number of chromosomes of other cells. When two gametes unite in a process known as fertilization, a diploid *zygote* forms, and the incredible process of development begins.

In this unit you will begin with the process of gametogenesis (spermatogenesis and oogenesis). Then you will follow the early processes of human development—fertilization and implantation. The unit concludes with a broad overview of fetal development and the anatomy of the fetal cardiovascular system.

EXERCISE 1: Meiosis

As you may recall from Unit 4, somatic cells divide by a process called mitosis. During mitosis, a cell replicates its genetic material and doles out this material into two daughter cells (see Pre-Lab Exercise 4 for a review of the stages). Each new diploid cell is identical to the original cell. During *gametogenesis* (oogenesis and spermatogenesis), however, we have two potential problems. For one, if each gamete were to have the same genetic material, we all would be genetically identical to our siblings, greatly reducing genetic diversity. Additionally, if each gamete were to have two sets of chromosomes, our offspring would have four sets and we would be tetraploid! To solve these problems, gametes undergo a process known as *meiosis* rather than mitosis.

In meiosis, also known as *reduction division*, gametes proceed through two rounds of cell division, with each resulting gamete having only one set of chromosomes (a *haploid* cell). In addition, in the process of meiosis, homologous chromosomes exchange pieces of genetic material, therefore enhancing genetic diversity. Meiosis begins in a manner similar to mitosis: The chromosomes replicate, and for a brief period the gametes have twice the normal genetic material (96 chromosomes). The cell then goes through meiosis I, which is similar to mitosis (see Pre-Lab Exercise 4 for a review of mitosis). But one important difference between meiosis I and mitosis occurs during prophase I. During prophase I, replicated chromosomes line up next to one another so closely that they overlap in several places—a phenomenon called *synapsis*.

Because each synapse has four chromosomes, the entire structure is called a *tetrad*. As the tetrads align themselves on the equator of the cell and prepare to attach to spindle fibers, the areas of the chromosomes that overlap form points called *chiasmata* or *crossover*. As anaphase I begins, the chromo-

somes exchange pieces of genetic material at the points of crossover.

Once telophase I and cytokinesis are complete, the gametes still have 46 chromosomes. They then go through the second round of division, meiosis II. Meiosis II is similar to meiosis I with one major exception: The genetic material does *not* replicate prior to the start of prophase II. At the end of telophase II, each gamete has only 23 chromosomes and is a haploid cell.

Modeling Meiosis

1. Obtain a set of meiosis and mitosis models.

2. First arrange the mitosis models in the correct order (see Figure 2.23 in your lab atlas for reference).

3. Now arrange the meiosis models in the correct order.

4. Looking at the models, what differences can you see between meiosis and mitosis?

5. As an alternative procedure, make model chromosomes with pipe cleaners. Use yellow and green pipe cleaners for the chromosomes undergoing meiosis, and red and blue pipe cleaners for the chromosomes undergoing mitosis. Show how the chromosomes duplicate and divide in the two processes. In addition, show how the chromosomes undergoing meiosis form points of crossover.

✔ Check Your Understanding

1. What happens if one of the chromosomes doesn't separate fully during meiosis II?

2. Give an example of a common disorder caused by this phenomenon. What are the symptoms of this disorder?

2 **EXERCISE 2:**
Spermatogenesis and Oogenesis _____

Gametogenesis is actually two separate processes: *spermatogenesis*, which takes place in the seminiferous tubules of the testes, and *oogenesis*, which takes place in the ovary and uterine tube.

Spermatogenesis

Spermatogenesis (see Figure 2.20 in your lab atlas) begins with stem cells located at the outer edge of the seminiferous tubules called *spermatogonia*. Before puberty, these diploid cells undergo repeated rounds of mitosis to increase their numbers. As puberty begins, each spermatogonium undergoing mitosis gives rise to two cells: one cell that stays a spermatogonium, and another that becomes a *primary spermatocyte*.

The primary spermatocyte then begins meiosis, dividing into two *secondary spermatocytes*, which migrate closer to the lumen of the tubule. The two secondary spermatocytes then undergo meiosis II, each giving rise to two haploid *spermatids*. Recall from Unit 27 that the spermatids then move to the epididymis to mature into functional gametes.

Most of the stages of spermatogenesis are identifiable on a microscope slide of the seminiferous tubules. In Unit 27 you examined the slide of the testis and noted the rete testis, Leydig cells, and seminiferous tubules. We will reexamine that slide here, except that this time we will focus on the seminiferous tubules to examine the stages of spermatogenesis. Because the seminiferous tubules do not contain functional spermatozoa, we also will examine a slide of a preparation of sperm cells.

Obtain a prepared slide of the testis and move to high power. Identify the *spermatogonia* on the outer edge of the tubule. They appear as cuboidal cells with round, centrally located nuclei. As you move closer to the lumen, look for cells with larger, more prominent nuclei. These are the *primary spermatocytes*. Why do you think the nuclei are larger in these cells?

On the inner edge of the tubule, you will see small, round cells with little cytoplasm. These are the *spermatids*.

Note that the spermatids lack the characteristic organelle seen in mature sperm—the flagellum. To see the flagellum, you will have to examine a slide of a sperm preparation. Because sperm are very small cells, the slide often has a thread or some other marker to let you know where to look. You will have to use high power, and if your lab has an oil-immersion microscope, you may want to use it. As you find the sperm, look for the *head*, in which the DNA resides, the *midpiece*, which contains an axomere and mitochondria, and the *flagellum*.

As you examine the slides of the seminiferous tubules and the sperm smear, use colored pencils to draw what you see, and label your drawings with the terms indicated.

Seminiferous Tubules
(See Figures 17.8–17.10 in your lab atlas.)
1. Spermatogonia
2. Primary spermatocytes
3. Spermatids

Sperm
1. Head
2. Midpiece
3. Flagellum

Oogenesis

Like spermatogenesis, oogenesis proceeds through meiosis to get a haploid gamete, the ovum (see Figure 2.21 in your lab atlas). But the two processes differ in some notable ways:

● *The early events of oogenesis occur during the fetal period.* Stem cells called *oogonia* undergo mitosis to increase their numbers to about 500,000 to 700,000. These cells then become encased in a *primordial follicle*, enlarge, and become *primary oocytes*. The primary oocytes begin prophase I, and then get arrested. Meiosis I does not resume until puberty.

● *The first meiotic division results in one secondary oocyte and one polar body.* At puberty, one or more primary oocytes are selected to complete meiosis I. The result of this process is a secondary oocyte and a small bundle of nuclear material called a *polar body*. The formation of a polar body instead of two secondary oocytes allows the oocyte to conserve cytoplasm, which it will need to sustain the cell.

● *Meiosis II completes only if fertilization takes place.* The secondary oocyte begins meiosis II before it is released during ovulation but completes the process, forming an ovum and a second polar body, only if fertilization occurs. If fertilization does not occur, the secondary oocyte degenerates without ever completing meiosis II.

The ovary provides a nice opportunity to view oogenesis and meiosis in action. Though you typically cannot see the chromosomes of the oocytes, you can determine the stage of the oocyte by looking at the surrounding follicle: Primary oocytes are encased in primordial and primary follicles, and secondary oocytes in secondary and Graafian follicles.

As you examine the slide of the ovary, draw what you see, and label your drawing with the terms below. Please keep in mind that one slide may not show all follicular stages, and looking at more than one slide may be necessary.

Structures of the Ovary

(See Figures 17.28–17.31 in your lab atlas.)

1. Primordial follicle
2. Primary follicle
   a. Primary oocyte
3. Secondary follicle
4. Graafian (vesicular) follicle
   a. Secondary oocyte
   b. Antrum
5. Corpus luteum

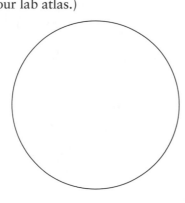

3 EXERCISE 3: Fertilization and Implantation _____

As sperm encounter the secondary oocyte, they begin a process called the *acrosomal reaction*, in which the acrosomes at the head of the sperm undergo *exocytosis* to release their digestive enzymes. The acrosomal enzymes digest the outer ring of granulosa cells that covers the oocyte, called the *corona radiata*. When the corona radiata has been sufficiently cleared, a sperm comes into contact with the zona pellucida and releases its enzymes to digest a path through this.

When the zona has been cleared, the sperm head and midpiece enter the oocyte but the mitochondria in the midpiece are destroyed (which is why only maternal mitochondrial DNA is passed on to the offspring). As the sperm

enters, the secondary oocyte finishes meiosis II to form an ovum and a second polar body, and further sperm are prevented from entering. The nuclei then swell to become *pronuclei*, and the nuclear membranes rupture. The two sets of chromosomes then fuse, forming a *zygote*.

After about the first 30 hours, as the zygote travels down the uterine tube, it begins a process called *cleavage*, in which it undergoes a series of mitotic divisions to produce two cells, then four, and so on.

At about day 3, it reaches the uterus and is at a 16-cell stage called a *morula* (which means "little mulberry"). The morula, which continues to divide, floats around the uterus for another 2 to 3 days until it becomes a hollow sphere called a *blastocyst*. The blastocyst has two populations of cells: a rounded inner cell mass called the *embryoblast*, and an outer layer of cells called the *trophoblast*.

At about day 6, the blastocyst adheres to the endometrium, and begins the process of *implantation*. The implanting blastocyst secretes digestive enzymes to burrow into the endometrial lining. This creates an inflammatory response in the endometrium, and the endometrium reacts to the injury by growing over and enclosing the blastocyst. The process of implantation is generally complete by the second week, about the time a woman's menstrual period would begin.

Time to Trace!

For this exercise, you will be tracing the gametes from their production in the male and female gonads to the point at which they meet to form a zygote, and finally trace the zygote's development to its implantation into the endometrium. As you trace, you may want to review Pre-Lab Exercises 1 and 3, Figure 18.1 in your lab atlas, and the Unit 27 exercises on anatomy.

Step 1: Male Gamete

Trace the male gamete from its earliest stage—the spermatogonium in the seminiferous tubule—through its stages of development to a mature sperm, and then through the male reproductive tract until it exits the body and enters the uterine tube of the female reproductive tract.

Start: Spermatogonium

End: Uterine tube

Step 2: Female Gamete

Trace the female gamete, the ovum, from its earliest stage, the oogonium in a primordial follicle, to the time at which it is ovulated and enters the uterine tube. See Figure 18.1 in your lab atlas for reference.

Start: Oogonium

End: Uterine tube

Step 3: Fertilization and Implantation

Trace the sperm and the ovum from the time they meet in the uterine tube to become a zygote, and through the stages of development until implantation occurs in the endometrium. See Figure 18.1 in your atlas for reference.

Start: Sperm and ovum meet in uterine tube

End: Implanted blastocyst

4 EXERCISE 4:
Development and Cardiovascular Anatomy __

The developing offspring, called a *conceptus*, undergoes multiple changes during the period of time known as *gestation*. Gestation is considered to be a 40-week period extending from the last menstrual period before conception to birth. The changes that the conceptus undergoes can be divided into three processes: (1) an increase in cell number, (2) cellular differentiation, and (3) the development of organ systems. This exercise asks you to examine these processes,

as well as the specialized development of the fetal cardiovascular system.

Development

As the blastocyst implants, it begins a process known as *embryogenesis*. During embryogenesis the embryoblast differentiates into the three primary germ layers—*ectoderm*, *mesoderm*, and *endoderm*. Two weeks after conception, when the three primary germ layers have formed, it is considered an *embryo*.

The next 6 weeks of the embryonic period are marked by development of the *placenta*, formation of the extraembryonic membranes, and differentiation of the germ layers into rudimentary organ systems. The placenta begins to form at day 11, when the trophoblast forms the *chorion*, the fetal part of the placenta. The chorion develops elaborate projections called the *chorionic villi*, which eat into uterine blood vessels to create a space filled with maternal blood called the *placental sinus*. Nutrients and oxygen from maternal blood diffuse from the placental sinus to the chorionic villi and are delivered to the embryo by the large *umbilical vein*. Wastes are drained from the embryo via the paired *umbilical arteries*. All three vessels travel through the *umbilical cord*.

During this period the extraembryonic membranes develop. One membrane of note, the *amnion*, completely surrounds the embryo (and later the fetus) and suspends it in *amniotic fluid*. The amniotic cavity protects the embryo from trauma by allowing it to remain buoyant in the amniotic fluid and protecting it from fluctuations in temperature. In addition, the cavity provides the embryo with ample space in which to move, which is critical to early muscle development.

As the placenta and extraembryonic membranes form, the primitive organ systems are laid down. This is complete by end of the 8th week, from which point the conceptus is considered a *fetus*. For the duration of gestation, the organ systems become progressively more specialized and the fetus continues to grow and develop.

Fetal Cardiovascular Anatomy

The cardiovascular system differs significantly in the fetus and the neonate. Within the fetal cardiovascular system are three *shunts* that bypass the relatively inactive lungs and liver and reroute the blood to other, more metabolically active organs.

1. *Ductus venosus*: The ductus venosus is a shunt that bypasses the liver. The umbilical vein, which brings oxygen and nutrient-rich blood into the fetus, delivers a small amount of blood to the liver but sends the majority of the blood through the ductus venosus to the fetal inferior vena cava.

2. *Foramen ovale*: The foramen ovale is a hole in the interatrial septum that shunts blood from the right atrium to the left atrium with the intent of bypassing the collapsed fetal lungs. Recall that the foramen ovale closes about

the time of birth, leaving a permanent indentation in the interatrial septum called the *fossa ovalis*.

3. *Ductus arteriosus*: The ductus arteriosus, which also bypasses the pulmonary circulation, is a vascular bridge between the pulmonary artery and the aorta. It also closes around the time of birth, leaving behind a remnant called the *ligamentum arteriosum*.

Identify each of the following structures on anatomical models or diagrams. See Pre-Lab Exercise 2 for reference.

Fetal Structures and Anatomy

(See Figures 18.1–18.5 in your atlas.)

1. Stages of development
   a. Zygote
   b. Morula
   c. Blastocyst
   d. Implanted blastocyst
   e. Embryo
   f. Fetus
2. Fetal membranes
   a. Amnion
      (1) Amniotic cavity
      (2) Amniotic fluid
   b. Chorion
      (1) Chorionic villi
3. Vascular structures
   a. Placenta
   b. Umbilical cord
      (1) Umbilical arteries
      (2) Umbilical vein
   c. Foramen ovale
   d. Ductus venosus
   e. Ductus arteriosus

Model Inventory

As you examine the anatomical models and diagrams in lab, list them on the inventory below and detail which structures you are able to locate on each model.

MODEL / DIAGRAM	STRUCTURES IDENTIFIED

✔ Check Your Understanding

1. Predict the effects if the foramen ovale or the ductus arteriosus fails to close shortly after birth.

2. Taking certain medications, such as nonsteroidal anti-inflammatories (e.g., ibuprofen), in the 7th to 9th month of pregnancy increases the risk for premature closure of the ductus arteriosus. What problems might this potentially cause for the fetus and the neonate?

3. A condition called *placenta previa* occurs when the placenta is too low in the uterus. Why do you think this is a cause for concern?

4. Active labor generally is considered to begin about the time that the "water breaks." What is the "water," and which membranes have ruptured to release it? What could be the potential consequences if the membranes were to rupture too early?

29. Physiologic Connections

■ OBJECTIVES

Once you have completed this unit, you should be able to:

1. Find physiologic connections between different organs and systems in the body.

2. Review the physiology of the organ systems that we have covered.

■ MATERIALS

- Anatomical models: human torso
- Laminated outline of human body
- Water-soluble marking pens

Often, in anatomy and physiology it is easy to see the anatomical connections. All organs are connected by the cardiovascular system, bones are united by ligaments and cartilage, and accessory organs of the digestive system are joined to the alimentary canal by common ducts. What becomes difficult to see is the *physiologic* connections that exist between all organs and organ systems.

At this point in our study of anatomy and physiology, we have examined each organ system and its individual organs and have studied how the parts fit together to make the whole. Along the way, we also have developed an appreciation for how the form of the system, organ, tissue, and cell all follow the primary functions of each part. Now, in this last exercise, we will develop an appreciation for how the physiologic functions of each organ complement one another.

Your final task in this lab manual is to find physiologic connections between seemingly unrelated groups of organs. It is easy to find the physiologic connection between the salivary glands, stomach, and duodenum, but the connection between the heart, bone, and thyroid may be a bit less obvious. Following are some hints, and an example, to make this task easier.

- Begin each exercise by listing the physiologic functions of each organ listed. Some functions are obvious, and others may require a bit of digging on your part. Take, for example, the skin: The skin has the obvious functions of protection and thermoregulation. But don't forget about its other tasks: The skin also synthesizes vitamin D when exposed to certain wavelengths of ultraviolet radiation, excretes small amounts of metabolic wastes, and serves immunologic functions. These less-discussed functions are easy to overlook but vitally important.

- Beware of the tendency to fall back to anatomical connections. When given the combination "lungs, kidneys, and blood vessels," it is tempting to say that the lungs breathe in oxygen, which is delivered by the blood vessels to the kidneys. That's all true, but this is actually an anatomical connection only. With some creative thinking on your part, you could come up with something much better. Remember that lung tissue makes the angiotensin-converting enzyme, and, having said that, the connection between the blood vessels and the kidneys becomes more obvious (well, at least it's obvious to me!).

- When in doubt, look to the endocrine system. Remember that many different organs, not just the organs discussed in the endocrine unit, make hormones. The heart, kidneys, and even adipose tissue all function in some respects as endocrine organs. For the organs that have a primary endocrine function, don't overlook some of the hormones

they make. For example, the thyroid gland makes T3 and thyroxine but also calcitonin, about which students tend to forget.

Here's an example of physiologic connection:

Lungs, Kidney, and Bone

Connection: The lungs have the function of gas exchange. When this exchange is inadequate, and oxygen intake is insufficient, the kidneys detect the decreased partial pressure of oxygen in the blood. The kidneys, in turn, make the hormone *erythropoietin*, which acts on the bone marrow to stimulate increased *hematopoiesis* and an increased rate of red blood cell maturation.

Wasn't that easy?

Now it's your turn.

Part 1: Heart, Parathyroid Glands, Small Intestine

Hint: Think electrophysiology of the heart.

Connection

Part 2: Testes, Bone, Muscle

Hint: What ion is required for muscle contraction to occur?

Connection

Part 4: Adrenal Glands, Blood Vessels, Kidney

Hint: What do the different regions of the adrenal gland make?

Connection

Part 3: Spleen, Liver, Large Intestine

Hint: Think about the functions of the spleen with respect to red blood cells.

Connection

Part 5: Pancreas, Sympathetic Nervous System, Liver

You're on your own for this one!!

Connection

Congratulations! You've done what you probably thought was impossible—completed this manual! I truly hope it has been fun (at least a little!) and that you have enjoyed it as much as I have. Best wishes with your future plans, down whichever path they may take you!

Index